GW00506043

A SKEW IN EVOLUTION

David Howells

A SKEW IN EVOLUTION
Copyright © DAVID HOWELLS 2012

First published by Zeus Publications 2012
http://www.zeus-publications.com
P.O. Box 2554
Burleigh M.D.C.
QLD 4220
Australia.

The National Library of Australia Cataloguing-in-Publication

Author: Howells, David.

Title: A skew in evolution.

ISBN: 978-1-921919-12-1 (pbk.)

Subjects: Evolution.

Human beings--Origin.

Zoology.

Dewey Number: 573.2

This book is a work of non-fiction.

The author asserts his moral rights.

© Cover Design—Zeus Publications 2012
Proudly printed in Australia

To Lyn and Claire, with love

Acknowledgements

The data in this book come from the internet, from the books listed in the bibliography and many others, from TV documentaries, magazine articles etc. The ideas are my own but they have obviously been stimulated by all of these plus hundreds of conversations, discussions and arguments with many people.

I cannot name every person who has influenced this work and I don't know whether they would all like to be associated with it, but I thank them all. It is often views that you disagree with most strongly that make you think most deeply about your own point of view.

There is one person, however, who I must acknowledge specifically – my wife, Lyn, who has suffered with this book almost as much as I have: for your patience in putting up with all this and much more, thank you.

Author bio.

David Howells was born in 1953 in Bedford, UK but grew up mostly in Bristol. He studied languages at Birmingham University but somehow ended up working as a systems analyst.

He moved to Australia in 1981 and now regards himself as Australian in everything but cricket (forever English) and rugby union (proudly Welsh). He lives with his wife and daughter on the Central Coast of New South Wales. This is his first book.

The Master looks deep into the scope. The universe twinkles and sparkles but his attention is focused on a little bluish planet circling a yellow star towards the bottom edge where a stain is spreading. He brings in into sharp definition. He looks up from the scope and turns to the Pupil at his side.

"Do you know what that is?" he asks.

"No, sir," replies the Pupil. "It started a while ago. Everything was looking great but then I noticed just a couple of them and now they're everywhere."

"What have you tried?"

"Well, I tried flooding the planet but a couple of them escaped and started up again. I tried sending in a special mutant of my own design but..."

The Master interrupts "...they killed it. And now they're out of control."

"How did you know?" asks the Pupil.

"You've got Man," says the Master. "You'll have to terminate the experiment."

"But..." begins the Pupil.

"I'm sorry." The Master smiles sympathetically. "It's a danger in any creation. Just a degree or two off, a bit too much carbon, any one of a hundred small errors and it can happen. Don't be down-hearted. It's your first try and the rest is very good. But there's no way round it – the experiment must be terminated."

"But surely there is something…"

"I'm sorry, but no," the Master's tone brooks no argument. "That's it – there's no cure. You must end it now or it will just spread and infect everything else."

The Pupil sighs. It's a shame; it was such a nice creation. But he knows in his heart that the Master is right – it has to go. He reaches for the red button but hesitates – perhaps one last look.

He peers into the scope but he can see it is hopeless. The pretty little blue planet is no longer quite so pretty; the once clear atmosphere has a dirty tinge; the blue waters have an oily sheen; the lush greens of the countryside are shrinking as brown and grey expand. He sighs again and his hand reaches towards the red button.

Meanwhile, on an unsuspecting planet...

CONTENTS

Introduction

When you consider nature, it is hard not to be impressed by its intricacy and its harmony.

Insects rely on flowers for food and flowers rely on insects for pollination. Some birds rely on trees for a meal of fruit, but the trees equally rely on the birds as agents to distribute their seeds in their droppings. Water evaporating from the oceans forms the clouds which fall as rain on the mountains and flow into the rivers which water the plains and thence back to the sea. The energy of the sun is synthesised by the grass of the plains that feeds the gnu and the zebra that provide a meal for the lion and the leopard and the hyena and the vulture who provide fertiliser that feeds the grass. There is a delicate balance between foodstuff, predator and prey. There is indeed a 'circle of life' in which every plant and animal has its place.

Every animal except one.

Ever since Darwin proposed that Man evolved from an ape-like ancestor, there has been debate as to whether Man is simply a rather sophisticated animal or is something more: this is a question of philosophy rather than of science. Darwin was convinced that 'the difference in mind between man and the higher animal, great as it is, is certainly one of degree and not of kind'[1]. I disagree. In a nutshell, my argument is that Man is an aberration, a skew in evolution. He is so far outside the mainstream that he can be regarded as unique. He is a brave experiment, the success or failure of which it is too early to judge.

Man dominates the globe from pole to pole and from sunrise to sunrise. He is by far the most numerous and is generally regarded as the most successful species on the planet today. But perhaps he is not. Somewhere in the transition to humanity, something was lost – that harmony with the environment that is intrinsic to the rest of the animal world.

Man is different from all other animals. There is a real and obvious distinction between 'natural' and 'man-made'. This is not because Man is not natural – he is not some cosmic interloper or a child of the gods – he developed from what came before the same as any other animal. But somewhere along the way Man went off on his own tangent. He became 'intelligent', and with intelligence came profound changes.

Man does not fit as well into the world as, say, an earthworm. He does not live within its limits and constraints. If it is too cold he heats it up, if it is too hot he cools it down. No other animal does this or could do this.

Man's ingenuity has allowed him to take over the whole planet and treat it as his own and he has done so ruthlessly. If he wants something he takes it. If something gets in his way he eliminates it.

---

[1] *The Origin of Species* by Charles Darwin.
In general I will not use footnotes, I will simply digress in the text if I consider it justified. Most references mentioned in the text can be found on the internet and the information available will usually be more comprehensive and up to date than anything I can quote. I have included a bibliography containing some of the texts used in the compilation of this work.

The result is what you see around you today. We dominate the globe but we are rapidly using up its resources, polluting its atmosphere, raping the land and slaughtering the animals. We have weapons enough to destroy the world several times over. The long-prophesied global warming now seems to be an imminent reality. Population continues to expand and push back the boundaries of wilderness. We are a species of seemingly infinite greed on a planet of finite resources.

Compared to other animals, Man has no 'nature'. A bird builds a nest by its nature as a bird – a crow will build a nest like a crow and a weaver bird will build a nest like a weaver bird. Each crow's nest is unique yet similar. If a human builds a house it may be anything from a shanty to a palace, a yurt to an igloo, a mud hut to a Manhattan penthouse. There is no instinctive, natural way to build a house.

If lion cubs play they play like lion cubs – they pounce, they rough and tumble. Similarly foxes or chimpanzees or any other animal that plays has certain set play behaviour. Not so Man. Human children may play tag or football or computer games or dominoes or any of millions of formal and informal games. There is no 'natural' play behaviour. What is the nature of Man? What is his natural diet, habitat, lifestyle? The questions are meaningless. Man has moved beyond nature.

Looking for the roots of human behaviour in the natural world is a futile search. While we have undoubtedly inherited many of the physical characteristics of our ape ancestors our brains and behaviours are for the most part, our own invention. Scientists point to the social nature of the chimpanzee or baboon lifestyle as being at the basis of human sociability. There is probably as much truth in this as in seeing the roots of human language in the alarm calls of certain monkeys. Yes, the social grouping of chimps may be a necessary precursory phase in the journey to the mega-societies of Man but Man has taken this and transformed it into something entirely new and radically different.

We look at bird migration or at the way a salmon or eel or turtle will find its way through the ocean to the place of its birth and we are amazed. We have invented many explanations as to how animals can do such things and why. Many birds migrate naturally as do salmon, eels, turtles, gnus, monarch butterflies and many other animals. They are in tune with the world, with its rhythms, patterns and seasons and when it is time to migrate they just go where they have to go. We are out of tune and we don't know where to go and so we have invented latitude and longitude, dead reckoning and global positioning systems. We understand things on our intellectual plane and then try to apply this intellectual understanding to nature. When we interpret nature through the distorting lens of our intelligence we are looking at things in human terms which may not be applicable to our subject. If a weaver bird ties a knot it does so naturally, instinctively. If a human ties a knot there is no nature or instinct involved, it is something learned and understood on an intellectual level. The two actions may end up with the same result but they are not really the same action and to treat them as if they are is mistaken.

Self-consciousness is the key to the problem. If an animal is not conscious of itself as an individual then it may regard itself simply as part of a larger whole. Thus an animal may think of itself (in as much as it thinks of itself at all) as an undifferentiated part of the herd or flock or shoal or simply as a part of the environment with which it interacts.

Man's self-awareness is the antithesis of this. He sees himself an individual, separate and alone. He is no longer a part of the world, he is apart from it and as such he is free to exploit it as he will. He has no relationship to the other occupants of the world and no sympathy with them – they are simply other, separated, individualised objects of his consciousness.

6

Man is a latecomer to the world. Modern Man, *Homo sapiens*, has been around in his current form for only one or two hundred thousand years. This may seem like a long time but in evolutionary terms it is barely noticeable, the blink of an eye. To give some sort of perspective, the dinosaurs dominated the Earth for some 200,000,000 years – we are not even in the ballpark on that scale. In the whole history of the world, millions of species have come and gone. The average life span of a species is somewhere around four million years; for large species this drops to perhaps a million years – that's still a good deal longer than we have been around. We have only just started. We have barely been around for as long as the Neanderthals we supplanted.

Human evolution has been astoundingly rapid. The rate of increase in brain capacity from around 400cc to our current 1,300cc is without precedent in evolutionary terms. In this aspect too we differ from all other animals whose evolution in general follows the gentler rhythm predicted by Darwin.

Homo sapiens has by no means proved an evolutionary success yet. Perhaps in a million years if we are still around then we can say we are successful but at the moment our fate is in the balance. Whether we will turn out to be an evolutionary success story like the dinosaurs or whether we will just be another failed experiment in evolution to briefly blaze and die, leaving no trace except an enigmatic extinction in the fossil record: this is a question for future history. At the moment we have as great a potential for spectacular failure as for success. Perhaps we are, to borrow Robert Goldschmidt's evocative term, "hopeful monsters" but whether we are more hopeful or more monstrous is still to be seen.

Uniquely, the success or failure of Man as a species is to a great extent in our own hands. We have the intelligence to understand our situation. We have the ability to regulate ourselves. We have shown almost limitless invention to get to where we are now. Do we have the intelligence, ability, invention and, above all, the resolve to overcome our rapacity and ensure a long-term future? That is a question only time will answer.

Scientific Truth:

Time is a recurring theme in this book so it behooves us to consider the time scale we are talking about.

The Earth is about four and a half billion years old, give or take half a billion years. The imprecision is deliberate. This is the current estimate and estimates tend to change.

Up until the early nineteenth century the biblical story of Genesis was generally accepted as literally true. In 1654, theologian James Ussher calculated the age of the Earth from biblical sources, and it turns out to have been created at nine a.m. on the 23rd October 4004BC. I suppose we may assume that Man arrived six days later on the 29th. This was the accepted orthodoxy for well over a hundred years. Fossils were either taken to be bones of dragons which must have been killed in the biblical flood or were simply not recognised as being evidence of earlier life forms. The first Neanderthal skull found in the 1850s was assumed to be from a slightly degenerate but contemporary village idiot.

With new discoveries, new techniques for aging and, most importantly, new acceptance of the great age of the Earth, the date of 4004BC is now looked on as a quaint mistake. But before dismissing Ussher it is well to remember that his figure was not just the wild guess of a misguided eccentric, it was a carefully calculated, exact figure derived from close study of the best source then available – the Bible. With no other evidence to go on, why should anyone dispute his figure? There was no carbon-dating, no way of scientifically establishing the age of the Earth. Why should anyone doubt that the world had been here forever; for thousands of years? That is a long time. Nobody thought in terms of millions of years in those days, let alone billions of years.

As scientific study and theorising went on, the biblical figure began to be questioned but it was not until the end of the eighteenth century when geologist James Hutton suggested that sedimentary rocks such as the famous White Cliffs of Dover, formed from the chalky remains of millions upon millions of tiny shells, could not possibly have formed so quickly that it was generally accepted that the world must be much, much older.

Just how much older was a matter of debate over the next century and a half. Counting tree rings and checking evolutionary progress throughout the geological record along with calculating rates of sedimentation and estimating heat loss, yielded results ranging from about 20,000,000 years down to about 500,000 years, with most leading scientists favouring a lower figure. This was the age of the Earth from the mid-nineteen hundreds, (around the age Darwin would have assumed), right up to the early part of last century when physicist Ernest Rutherford discovered atomic disintegration.

Atomic disintegration is the change of a more complex form (isotope) of an atom into a simpler form over time. Rutherford calculated that the energy loss in this action would account for all the heat lost from the Earth since it was created and the Earth suddenly aged considerably. The rate of this 'isotopic decay' is regular and measurable and is used in many cases to establish the age of fossils and rocks.

It is only about fifty years ago that the current figure of 4,500,000,000 years old came into general acceptance. This figure has gone up or down by 500,000,000 or so since it was first postulated and it continues to oscillate with ever more discoveries and ever better aging techniques. It may yet turn out to be another quaint mistake if, for example, it is discovered that isotopes don't always decay at a reliable rate under all circumstances.

There are two reasons for the above story:

The first is because it is only fair to say that many facts and figures given in this book are the best I have at this time but are subject to review and should not be taken as gospel – any more than should the story of Genesis. The facts and figures I use in this book – although, I must emphasise, not the opinions I express – are, to the best of my ability, an average consensus of the many different sources I have researched.

However, the more you research in many of these areas the less consensus you find. In discussing evolution and especially the origins and nature of Man, we need to be especially careful of the phrase 'scientific orthodoxy'. There is no absolute orthodoxy. New evidence, new techniques and new interpretations make this one area where there are almost as many theories as there are theorists. There are more anthropologists today studying more evidence in more detail than ever before. Such minutiae as microscopic scratches on millennia-old stones or chemical analysis of fossilised faeces tell the trained scientist a story of behaviour or diet or lifestyle.

There are areas where there is no agreed single scientific orthodoxy and even in those few areas where the facts are agreed, there are always details which are disputed. What is more, current theories are not forever.

In the thirteenth century, an educated European with an interest in astronomy would probably have followed the science of the ancient Greeks. The main source of knowledge regarding the structure of the universe came from the writings of Ptolemy (actually an Egyptian) who wrote or compiled several books on the subject in the second century AD. In the Ptolemaic universe the Earth was at the centre with the sun, moon, planets and stars circling it in spheres as God had planned.

By the nineteenth century most educated people would have believed in a Newtonian model of the universe. Our solar system had the sun at centre, the planets spinning around it in elliptical orbits and the stars were strewn across the faraway sky. This was still as God had planned.

Today most scientists see a universe still expanding from the Big Bang in four Einsteinian dimensions of space/time. The Earth is a not especially important planet in an ordinary galaxy tucked away on the edge of the universe and God simply doesn't rate a mention anymore.

But before dismissing the last few thousands of years of learning as so many false trails on the slow path to the truth, it is worth remembering that the old theories lasted much longer than any current theory has and they did not die easily. The Ptolemaic model of the universe was the accepted model from about 2,500 years ago until only about 600 years ago – a span of about 1,900 years. The current Big Bang model has been the accepted model for only about 60 years and new models lurk in every corner. What will be scientific orthodoxy tomorrow or in a hundred or five hundred years time is anybody's guess. My guess is that it will be radically different from today.

Time:

The second reason for the discussion of the age of the Earth above is because it introduces the problem of time scale which is a central theme of this book.

Most people live for less than one hundred years and many for much less than this. A thousand years is longer than we can truly imagine and a million and a billion[2] become almost meaningless other than being convenient labels for 'a long time' and 'an even

---

[2] I use the American billion throughout i.e. 1,000,000,000

longer time'. Even in writing there is a problem. The difference between 'a million years' and 'a billion years' is more than a single letter. The difference is in fact a billion years to an accuracy of 0.1%. Writing 1,000,000 and 1,000,000,000 gives no better impression – it is simply one and a few noughts or one and a few more noughts. In reading quickly it is very easy to confuse them. In some languages, instead of one, two, three, four etc., the numbering system goes something like one, a couple, several, many – we tend to think of time on a grand scale in the same way.

There is a further problem once we get to the smaller timescale of Man: the years ago/BC dilemma. In Australia where I am writing this book, this year is designated 2011AD. The AD stands for Anno Domini and refers to the (presumed) date of birth of Jesus of Nazareth. Dates prior to 0AD may be referred to as BC – Before Christ. It is very easy to confuse 3,000 years ago and 3,000BC which is in fact 5,000 years ago. To avoid this I will always use 'years ago' unless giving a specific date. Of course in the grand scheme of things a couple of thousand years is insignificant. If the question is: when did the dinosaurs die out? It makes little difference whether you say 65,000,000 years ago or 65,000,000BC.

To get a clearer grasp of timescales, the classic analogy is to compress the whole history of the Earth into one day. We start with the creation of the Earth at 00:00; life emerges with the dawn at around five a.m. By ten o'clock blue-green algae are common and, slowly, more complex life forms begin to develop but nothing much exciting happens until coming on for nine in the evening when there is an explosion of activity. Animals with hard body parts suddenly appear followed closely by fish and then amphibians. The dinosaurs have their reign between about ten thirty and eleven forty-five p.m. Mammals begin to dominate in the last fifteen minutes. At one minute to midnight the first hominids appear and finally *Homo sapiens* arrives just as the clock ticks over to midnight. This does give some idea of the relative longevity of Man – blink and you miss him – and some idea of scales.

Alternatively, if I wrote a 100-page book on the history of the world, starting with its creation on page 1 and with equal amounts of space dedicated to equal amounts of time, life would appear on about page 25. Another 63 pages pass before the Cambrian explosion on page 88. Amphibians appear on page 93 and then we have reptiles from 94 to 99. At the end of page 99 we meet mammals for the first time and in the last paragraph are the first hominids. In this scheme, *Homo sapiens* is literally the last word in evolution.

These analogies themselves highlight another common misconception: that Man is the epitome of evolution. The world has been around for a very long time. There is no reason to believe that it will not be around for a very long time to come – the current estimate is that the sun will last for about another five billion years and the Earth may last along with it. For the last three point eight billion years or so there has been life on Earth and this life has been evolving and there is no reason why life should not continue evolving in some form until near the last days of the Earth. We happen to be at a certain point in this continuum, a no more significant point than, say, one hundred million years ago when Man did not exist or one hundred million years hence, when Man is again unlikely to exist in anything like his current form. Evolution has not culminated in Man, he is not the splendid flower on the topmost branch of the evolutionary tree; he is simply one more twig on the bush of evolution. Whether that twig grows or dies and, if it grows then what it grows into are questions that only time can answer.

Because we are here and now, we view history in a distorted perspective. Recent events tend to loom large while the distant past fades to obscurity. Also, because we can remember the past but cannot foresee see the future, it is easy to think that the present is the end of the story instead of just a point along the way.

This perspective has led to the formulation of the 'anthropic principle'. In its weak form this states that Man could not exist if the world were any other way – bigger, smaller, closer to or further from the sun etc. This is trivially true. The strong anthropic principle conjectures that the world, the universe, indeed all of creation has been a build-up to the flowering of Man, specifically of Man's intelligence. The argument is basically that there are many, many circumstances which had to be just so for Man to have developed and for Man to have developed intelligence. What's more there are certain highly unlikely cosmic coincidences which seem to be waiting for Man to discover them. The combination of all these circumstances is so extremely unlikely that there must be some guiding principle behind them.

I have never been able to decide whether this strong anthropic principle is serious or tongue in cheek. If it is serious, I disagree. Exactly the same arguments could be used by a dog to prove that the universe was created just so that dogs exist. What is more, the same arguments apply to all denizens of all worlds at all times.

It has been said that creation myths tell us how a society sees itself in relation to the rest of creation. The story of this book is a contemporary creation myth. We may think we have abandoned gods and magic for science but in reality we have only swapped one type of magic for another. When scientists talk of the Big Bang or black holes or curved space-time they are modern shamans reciting the holy mantras.

This book is largely the story of the last second of our earthly day. It is based on today's scientific truth. This may or may not turn out to be better than yesterday's truth and it will almost certainly be superseded by tomorrow's. But be that as it may, we are here and we are now and so we must accept today's truth and hope that tomorrow does not make too great fools of us.

Man Versus...

Man has had an undeniable impact on this planet. He has changed the environment radically. But he is not the first force to do this nor will he be the last. There have been events in the past which have caused huge perturbations in the grand scheme of life. The very first living organisms irrevocably changed the Earth's atmosphere. This created an environment where life could take-off A few billion years later, a large meteor hit the Earth – bang go the dinosaurs, welcome the mammals. There have been Ice Ages and heat waves before, far more extreme than the global warming we are experiencing so far. There has been volcanism beyond anything Man has ever seen; massive rises and falls in sea level; droughts and floods beyond imagination. And through it all, life has continued and continued evolving.

To put things in some sort of perspective, between each chapter we will compare Man to some other phenomenon.

Man versus Climate

There is clear evidence that we are currently experiencing a period of global warming. There are growing holes in the ozone layer. There is increasing pollution of all types. There is no question that Man is contributing to these problems and has been for many years although the degree and impact of this contribution is still disputed. There have recently been conferences where these issues have been discussed and slowly we are coming to accept our responsibilities towards the planet. Exactly what our response should be is a matter of fierce debate and whether we actually have the political will to do anything effective about these responsibilities is another matter altogether.

One of the main problems is the massive amount of so-called greenhouse gases which have been released into the atmosphere in the last few hundred years since the Industrial Revolution. By polluting the atmosphere and preventing the escape of excess heat, these gases cause the Earth to warm up something like a greenhouse. The increase in greenhouse gases is exacerbated by the reckless destruction of forests which have in the past acted to mitigate the noxious effects of such pollution.

It is common to bemoan the current clearances in Amazonia but this sort of destruction has been going on for a very long time. The reason that places like the Amazon and South East Asia are taking such a battering is because other sources have been exhausted. In the Middle Ages and before, the forests of Europe were destroyed to create farmland and pasturage as well as for ship-building and general building. Thousands of trees were felled to make up the great merchant and military fleets of the sixteenth and seventeenth centuries. Pollution has an equally long or longer history with some evidence to suggest that salt build-up was a side effect of irrigation even in biblical times.

In messing with the biosphere we are playing with fire. The Earth is a huge feedback system. As elementary chaos theory tells us, in any feedback system there may be a point where gradual change can give way to sudden radical change.

Just these sorts of radical changes have happened in the past with the Ice Ages. In these events massive amounts of ice, kilometres thick covered much of the globe. Glaciers sliced out the fjords of modern Norway and New Zealand. The huge amount of water locked in ice caused sea levels to fall by hundreds of metres. The whole contour of the globe was changed – land bridges opened up, huge areas of the world became barren wasteland.

Man was around during the last Ice Age – the Wurm or Wisconsin glaciation – which ended only a short 10,000 years ago. Northern America was nearly completely covered by the huge Laurentide ice sheet. In Europe and Asia, the Scandinavian ice sheet reached the northern parts of the British Isles, Germany, Poland and Russia. Actually this event proved of major benefit to humans as the drop in sea levels allowed them access to North America and Australia.

How would Man stand up to an Ice Age now? We could not stop the ice – nothing could stop the ice. We could run to the temperate zone but the over-crowding would be horrendous and resources would be stretched beyond their limits – confrontation would be inevitable. We could adapt to the ice to some extent with warmer clothing and increased heating. We could adapt our agriculture.

We could go underground although only in relatively small numbers. No doubt billions would die. But Man survived, indeed flourished, in the last Ice Age and there is little reason to doubt that he would survive a new one, along with those others he chose to let live.

What of global warming? Suppose the current trends continue and increase. The ice-caps and glaciers shrink and sea levels rise. Low lying islands are inundated. Polar bears become extinct. The Bay of Bengal floods. Huge numbers of people in coastal areas are displaced. Weather patterns become erratic – droughts and floods become common, huge storms crack over the Earth, tornadoes and hurricanes rip around the world. Man would adapt. Millions would die, perhaps billions, but Man would move to higher ground, change his crops and survive, probably much better than most other terrestrial species.

The devastation which would be caused by either an Ice Age or global warming would be tremendous. An ice sheet would totally raze any city in its way – all that would be left would be a moraine of concrete fragments. Similarly a super-storm could destroy a city in minutes and rising sea-levels could simply erase coastal settlements without a trace. We have certainly done similar things on a limited scale – Hiroshima and Dresden come to mind – and the current destruction of nature is perhaps comparable, but against the full force of nature, Man's works sink to the level of trivia.

In The Beginning…

In the beginning was a singularity.
And the singularity went BANG!!!

The world as we experience it today was not always this way. According to the most popular scientific theory of today, the universe started a very, very, very long time ago with a Big Bang. What was there before the Big Bang (if anything) and what caused the Big Bang (if anything) are questions that orthodox science cannot answer. Life, the universe and everything including time began with the Big Bang.

The Big Bang was big. At the time of the 'Bang', which by current estimates was between fourteen and twenty billion years ago, the whole universe was compressed into virtually nothing, a singularity. This was not merely microscopically small, it was smaller than that. It was a sphere (though perhaps not a perfect sphere) with a radius of nearly zero: in the words of Armand Delsemme "a microscopic four-dimensional hypersphere, finite but unbounded". Then BANG!! One millionth of a second later the universe was several hundreds of millions of kilometres across with a temperature of several trillions of degrees.

The universe has never got over the Bang. It is still expanding today. Everything is still rushing away from the centre like debris from an explosion. Whether the debris will simply settle down or will keep expanding forever or will fall back on itself and explode all over again is one of the great mysteries.

How do we know all this? When a racing car speeds past you, the engine note rises and fall as it passes. This caused by distortion of sound waves coming and going and is called the Doppler Effect after the Austrian physicist, Christian Doppler who first noticed it or at least who first explained it. The important thing is that this difference between coming and going applies not only to sound but also to light. Light tends to shift from bluish to reddish, coming and going. Looking around the universe, everything is reddish which implies it is all moving apart and the further away you look the further the light is shifted to the red end of the spectrum. So if everything is moving apart then it must at one time have been closer together and been pushed apart by some force. Working backwards from here we get a singularity and a bang (the mathematics is actually quite complicated but there is no need to go into it here).

There are other, rival theories – Fred Hoyle's 'Steady State' for one – and any number of refinements and details, most of which are too subtle to be understood by non-physicists like me, but the Big Bang is the most widely accepted theory .

This theory of course relies on our current understanding of light and the forces which shape the universe and also makes the assumption that the forces which we believe shape the universe are the same now as they have always been. This may or may not be true. It is also worth remembering that the Big Bang theory, which is now accepted as 'truth' by practically the whole of the scientific community is less than one hundred years old.

After the Bang the universe was incredibly hot. In considering evolution we will meet some big figures but the figures involved at the early stages of creation are just mind-bogglingly big. There is little point in going into detail. There are many fine books on how the universe began and no doubt by the time you read this book, many more will have been published with fine new theories and even bigger numbers.

Anyway, the universe kept on expanding and dissipating some of this heat until it was cool enough for matter to form. What was there before there was matter? The bits and pieces that make up matter – hadrons, leptons, neutrons, electrons and so on – all floating around in a sort of cosmic plasma. There was probably a quark or two in there as well. We might learn a bit more once they start smashing particles in earnest at CERN's Large Hadron Collider, or perhaps we won't.

Once matter formed, gravity started to pull it into clumps. Some of these clumps started to collapse onto themselves under gravitational attraction and eventually the heat generated by this collapsing matter started new nuclear reactions and stars started to glow. These processes are still going on today. Our own galaxy, the Milky Way, was formed at some stage, and at the far end of one arm of this galaxy a yellow star began to shine. Around this star a set of planets formed.

The Earth is the third of these planets. The Earth formed between 5,000,000,000 and 4,000,000,000 years ago. For several hundred million years not a lot happened, then life began.

Life:

Before going further we probably should define 'life' but we shall not. Firstly the definition of 'life' is not easy: there are degrees of life as there are degrees of most things. And secondly the exact definition is not important. The main stipulation that needs to be made is that 'life' included a very clever, self-replicating molecule called ribonucleic acid. There are two main forms for this super-molecule – RNA (simple ribonucleic acid) and DNA (deoxyribonucleic acid). Which came first is a matter of lively debate but this is an academic question. The point about this DNA molecule is that it is now taken to be the engine of evolution. DNA is an extremely interesting molecule.

Firstly it is very big – a single molecule of human DNA if stretched out would be several centimetres long. This exceptional but even viral DNA which is much simpler is still gigantic by normal molecular standards. Of course, in its natural state, DNA is not stretched out in a line; it is contorted into a compact, complex unit only nanometres in diameter.

Secondly DNA has a unique structure – the famous double-helix – something like a twisted ladder. The 'ladder' has many, many rungs and each rung is made up of a pair of the chemical bases adenine (A), cytosine (C), guanine (G), and thymine (T). These bases are 'coded' so that each one will always pair with its partner base and with no other: guanine always pairs with cytosine and adenine always pairs with thymine.

Thirdly, DNA seems to be common to all living things, certainly all living things on Earth. The sequence of bases may be different in different animals and plants but all living entities have DNA and it all works the same way.

Now comes the really clever thing about DNA: the molecule can unzip right down the middle into two separate strands, each strand being made up of one of the base pairs. Each of these bases will then find a partner base to form a new rung and, lo and behold, we have two new DNA strands each one exactly the same as the original strand.

DNA comes in units known as chromosomes. In humans there are twenty-four chromosomes making up what is known as the human genome. If all the DNA in all these chromosomes was unravelled and laid end to end it would be about three metres long – that is about three billion base pairs.

What DNA does within an organism is to direct the synthesis of proteins. Seeing that our bodies and the bodies of all living things are made up of proteins this is a very significant function. The sequences of bases in the DNA strand cause the molecule to contort in a particular way and these contortions are instrumental in making one section of the molecule synthesise one particular protein and another section produce a different protein. What is more, the same sequence of bases synthesises the same protein no matter whether the DNA comes from a daisy, a dolphin or a diatom. Although the exact specification of the DNA of each species is unique, the basic structure and mechanism is the same no matter if we are looking at a bacterium or a human.

Of course there is a lot more to DNA than this simplistic overview but there is no need to get into further detail at this stage.

We as a species are now beginning to understand something of DNA. The whole sequence of some DNA molecules has been mapped including human DNA and even Neanderthal DNA. We can cut sequences from, say, the DNA of a cold-water fish and paste them into the DNA of, say, a strawberry to produce a strawberry which is frost-resistant. We can snip and link human DNA into pigs to get them to grow human-compatible transplant organs. We can create clones; that is produce genetically identical copies of an individual. So far this technology has been successfully applied to animals as large as sheep but has not yet been applied to humans, indeed human cloning is illegal in most countries which have the technology to do it. However the first step on the slippery slope to human cloning has been taken – cloning human stem cells. It seems inevitable that sooner rather than later someone somewhere will clone a complete human.

There is still much we do not know or understand. As mentioned above, sequences of DNA bases function to synthesise proteins. However in our current models most of the DNA string seems to be made up of 'noise' – meaningless sequences of bases which do not synthesise anything. These sequences are often referred to as 'junk' DNA. Perhaps these are simply junk or perhaps there is more to DNA that we have yet to discover. Time will tell.

Exactly how life began is unknown. There are several theories, probably the most popular being the 'primordial soup' theory. The theory goes that in the early days of the Earth there were extreme physical conditions – great heat, pressure, electrical storms etc., and that the early seas were a rich chemical broth. These seas bubbled and seethed with chemical activity – molecules formed and broke, joined and split, mixed and mingled until, presto, the right chemicals chanced upon each other and DNA was born. Of course it might not have been quite that simple. It has been suggested that certain chemical bonds are more stable than others and that these more stable combinations formed into molecules which 'evolved' a la Darwin to the eventual complexity of DNA.

In 1953, chemists Stanley Miller and Harold Urey ran an experiment where they bombarded a mixture of gases and water vapour with electric sparks. After leaving their experiment running for a while they found that complex chemicals had formed including some amino acids, the building blocks of life. This may well indicate that the 'primordial soup' theory is somewhere on the right lines but it is a quantum leap from producing a few amino acids to producing nucleic acid and from there it is another quantum leap to producing life. The Miller/Urey experiment has since been repeated with varying degrees of success although many of the molecules considered essential for life have never been produced. However, not knowing the exact conditions prevalent on Earth at the time that life appeared and not having several hundred million years to run the experiment, we are unlikely ever to synthesise life in the laboratory from scratch.

The recent invention of a new form of self-replicating DNA which was so sensationally hailed in the media as the "creation of life" was not actually creation of life from non-life. The new life form was created by dicing and splicing existing DNA into a hitherto unknown, stable life form. While this is a tremendous achievement it does not really address the origin of life.

Scientists are rushing ahead with genetic experimentation despite our incomplete knowledge and reservations about long-term safety and moral, ethical and legal questions. We are growing and eating GM (genetically modified) food crops regardless of the risk of man-made DNA getting loose into the natural world. We are engineering genetic cures for genetic diseases, making 'designer babies', working towards genetic immortality with as little knowledge of what we are playing with and as little regard for consequences as with early atomic testing. We are tugging at loose threads in the fabric of life: let us hope we don't unravel anything we cannot put back together.

One other major theory is that life originated somewhere else in the universe (or perhaps everywhere else in the universe) and came to Earth on meteors which were bombarding the Earth at that time. There is some evidence to support this in that organic material has been identified on some meteorites but it is far from certain that this is actually evidence of life. The problem with this theory is that it does not answer the question of the origin of life at all, it just dodges it. Life developed somewhere, somehow. If it was not on Earth three point eight billion years ago then we are still left with the questions: where, when and how?

The problem is that if we reject the 'primordial soup' theory then we have to propose something else and the only other serious contender is a god. This then leads to the 'if a god created life then who or what created god?' argument which is the epitome of the old saying that all arguments have two sides but some have no end. We take the age of the Earth to be somewhere around 4,000,000,000 or 5,000,000,000 years. The first glimmerings of life appear about 3,800,000,000 years ago. Our knowledge of this period is sparse as the only evidence comes from micro-fossils in some very old rocks and there are very few rocks this old with unambiguous fossils in them. What fossils there are show simple single-celled organisms with a cell wall but without a separate nucleus: these are called prokaryotes. Prokaryotes are about as simple as life gets but they still have a level or organisational complexity millions of times greater than the rocks they live on or the water they live in. Prokaryotes thrive in extreme conditions. They are still the most abundant life form on Earth.

The implication of the above is that life arose after a maximum of 'only' about 1,000,000,000 years and possibly within less than 500,000,000 years. It may well be that this happy event was an inevitable outcome of the conditions on the planet at that stage of its development and that therefore life is a common, indeed inevitable, event on other planets. On the other hand it could be that the coming together by chance of a molecule as complex as DNA is such a vanishingly small possibility that the invention of life is likely to be a unique event. Both views are plausible and which view you adhere to is a largely a matter of faith.

Whether there is life 'out there', what that life may be like and whether we will ever make contact are fine cocktail party conversations but not are not really questions likely to be answered in the near future. Astronomer Frank Drake has produced a neat little formula which calculates the number of civilisations in the universe with who could potentially communicate with us, which takes into account everything from the number of new stars created in a year to how long a civilisation with the ability to communicate with us would last. The actual formula is:

$$Nc = R * fp * ne * fl * fi * fc * L$$

Where Nc is the number of civilisations, R is the number of stars created each year, fp is the fraction of those stars having planets, ne is the number of these which are, like Earth, capable of supporting life, fl is the fraction of those planets on which life occurs, fi is the fraction of those life forms having intelligence, fc is the number of those intelligent life forms getting to a point where they could communicate over interstellar distances and L is the length of time that such a communicating life form may last.

The problem is that while we can make informed guesses at some of the parameters – the number of stars formed each year is perhaps ten and the fraction of these having planets is about one in ten by current guesstimates; in our solar system only the planet Earth is positioned at such a distance from the sun that life (as we know it) could develop and there is no reason to assume that this figure would be much different for other solar systems. Beyond that you may choose any figures you want. If you think that life is an inevitable consequence of the fundamental physics of the universe you could assign that the value of 1. If on the other hand you consider life to be a billion to one chance then you could assign it a value of .000000001. Of course you could equally well choose any value in between. Similarly, the fraction with intelligent life and the means to communicate are matters of complete conjecture. As for how long an advanced civilisation capable of inter-stellar communication might last – we do not have enough data to make any guess at that. We sent a craft into the far depths of space only a few years ago. Our interstellar communicating civilisation has so far lasted only a few scant decades. In terms of the formula above this might as well be zero.

If we assume that there are ten new stars a year and the other factors are about one in ten except that only one in hundred intelligent life forms will ever get the point of interstellar communication and if we assume our civilisation will last for 10,000 years (and that's about three times as long as Ancient Egypt lasted) then the formula is:
Nc = 10 * .1 *1 *.1 *.1 *.01 *10000

If you work this out it comes to 1. We are alone in the universe.

On the other hand, if we assume that there are a hundred new stars and that life is inevitable and will inevitably develop to a point where interstellar communication is possible and that civilisation will last as long as the species lasts which averages four million years then the formula is:

$$Nc = 100 * .1 * 1 * .1 * 1 * 1 * 4000000$$

If you work this out it comes to 4,000,000. We are one of many.

Be that as it may or may not, by whatever means and for whatever reason, life did begin on Earth and in the beginning it literally changed the world. One of the things the early life forms did was to process carbon dioxide into oxygen by photosynthesis just like modern plants. In doing so they created the oxygen atmosphere we enjoy today and set the stage for an evolutionary bonanza.

It took around another 1,000,000,000 years before the next major step – the development of eukaryotic cells with a separate nucleus and other bits and pieces in the cell – and then perhaps another 1,000,000,000 for the next great leap forward in evolution: the emergence of more complex, multi-cellular organisms. How cells acquired a separate nucleus, mitochondria, flagella etc., is a hot topic of debate. One of the most plausible suggestions is that these were originally independent organisms which united into a sort of symbiotic community. The development of multi-celled organisms can be viewed in much the same way. This fits well with a co-operative view of the world rather than a competitive, combative view – we will come back to this point.

Why it took so long for life to develop from single to multiple cells is difficult to explain – in retrospect it seems such an obvious idea for single-celled organisms to combine into more complex forms that you wonder why they did not get together earlier. But this does highlight one of the recurrent features of evolution – progression by sudden leaps and bounds. The era of no-life was relatively short – 1,000,000,000 years give or take a few 100,000,000. Then there is the leap from non-life to simple prokaryotic life. This was the only stuff around for about 1,000,000,000 years when things start to get a bit more complex and we get the beginnings of multi-cellular life forms. While these types develop, nothing much changes for about 1,500,000,000 years and then, suddenly, about 550,000,000 years ago, we have one of the greatest and most enigmatic of all leaps – the Cambrian explosion.

In the blink of a geological eye (about 35,000,000 years), many new life-forms arose, forms that were radically different from all that had come before and some forms that would last until modern times. Suddenly there were animals with hard body parts, shells and suchlike. In fact, all the major life-forms (phyla) appeared in the Cambrian Explosion and, even more remarkably, no new ones have evolved since.

Until relatively recently the only known fossils prior to this proliferation of forms were of single cell or very simple multi-cell animals which could not be the immediate precursors of the new animals but new fossil finds have been made dating back to almost 700,000,000 years ago. The problem is that many of these so-called Ediacaran fossils are of life forms which are quite different from the Cambrian forms, forms which appear to have completely disappeared. Far from shedding light on the mystery of how fully fledged life forms could arise with no apparent predecessor, the Ediacaran fossils have just raised new questions as to who were the predecessors of the Ediacaran forms and where are their descendants.

The current rationalisation of the Cambrian Explosion is that the ancestral forms did exist but were entirely soft bodied and so have left little or no fossil trace and this may be true. Most Pre-Cambrian rocks have been eroded to nothing or subsumed into the Earth's mantle or otherwise destroyed so rocks over 500,000,000 years old are rare and it is rarer still to find rocks this old with fossils in them. It is quite probable that we will never have unequivocal evidence of the precursors of Cambrian life but whether these intermediate forms exist or not, the fact is that by about 500,000,000 years ago we have the ancestral forms of animals still around today. Among these animals are sponges, jellyfish, starfish and flatworms.

Flatworms are the most interesting of these to us as they are our first recognisable forebear. Some flatworms such as tapeworms have now specialised in living parasitically but there are many others which still slime about much as they did 500,000,000 years ago. You may not think of yourself as very similar to a flatworm but you are. You have a head and a tail (sort of), a front and a back, a left and a right; you have a brain, a central nervous system and sensory organs such as a pair of eyes: flatworms have all of these. What is more, the very first flatworms had all of these.

But before continuing with the story of the development of life on Earth we must consider the forces which were, and presumably still are, directing that story.

Evolution:

In general the story of life is a story of increasing diversity and complexity. Where there used to be only a few species of simple life-forms there are now a plethora of complex animals and plants.

Our current model of the relationships among animals and plants is based to a large extent on the work of Carl Linné, better known by his adopted name of Carolus Linnaeus. A Swedish botanist born in 1707, Linnaeus published his *Systema Naturae* in 1735. This classified the living world into a hierarchy according to observable physical characteristics. While Linnaeus' classifications have been greatly expanded and modified since their original publication, several aspects remain constant: the hierarchical organisation, the use of physical characteristics as a basis of classification (although the definition of 'physical characteristics' has changed greatly) and the use of binomial nomenclature i.e. the genus and species name as a unique identifier for any living entity e.g. *Homo sapiens* as a designation for modern humans (genus *Homo*, species *sapiens).* Of course with the recognition of the significance of DNA and improved techniques of analysis, DNA is now one of the most significant physical characteristics used in classification.

Linnaeus' original hierarchy sorted the world into Kingdom, Phylum, Class, Order, Family, Genus and Species e.g. humans in this scheme are in the Kingdom of animals, the Phylum of chordates (having a spinal cord), the Class of mammals (having mammary glands), the Order of primates (having grasping hands and depth perception), the Family of hominids (bipedal, no tail), the Genus homo (big brained, tool-reliant), and the Species sapiens (self-aware). Nowadays there are finer divisions into sub-phyla, sub-classes, sub-orders, infra-orders and super-families but the principle remains the same.

The question of how and why all these different life forms come to be, how they are related and where or indeed if Man fits into this hierarchy is a question that has been asked down through the ages.

In every culture there is a creation myth to explain the world and Man's place in it. The creation story I know best is from the Bible book of Genesis. In brief, God split the Earth off from heaven; made night and day; made land and sea; made plants; made the sun and the moon; made fish and fowl; made animals; created Man, both male and female, in His own image. A while later there was a flood in which most of the world was destroyed but one righteous man and his family plus assorted animals survived and the descendants of these survivors are the inhabitants of the world today.

This is similar to the creation myths of many other cultures. In most creation myths we start off with nothingness – light or darkness or void – and then a god or gods, often male and female, create the world. Man is usually unique, created last and separately from other animals, often to be a companion of the gods or at least with a special relationship with them. In many cultures there is also a flood or disaster where the gods are displeased with their work, often specifically with Man, and set about cleansing the world. But there is always a Hero to start the good work again.

Up until relatively recently in those countries where Christianity and Islam dominated, the biblical creation story was accepted as literal truth or at least as a plausible explanation of the state of the world. In other parts of the globe there were other myths which were accepted as the orthodox view. It was assumed that the world was as it was because God had ordained it so and that it had always been that way since the flood.

By the mid nineteenth century the plausibility of biblical creation was becoming more and more strained. New ideas on the great age of the Earth, fossil finds, new evidence from geology and zoology all combined to spread doubt on the creationist view. The spirit of scientific rationalism was rife and this spirit could not be satisfied with anything other than a scientifically rational explanation. Gods and bible stories did not fit in this framework. The stage was set for a new evolutionary theory.

The idea of evolution itself, of one form transmogrifying into another over time, had been around for many years – the Greeks probably suspected it. The problem with evolution is how and why does it happen? Where do all these animals come from? How to account for the variety, the variations and the similarities, the actions and inter-actions?

Up to this point nobody had put up a convincing argument as to why these changes should occur and why they should lead to the world as we see it. It was assumed, as it generally still is, that evolution had progressed from simpler to more complex forms and that Man, being the most complex and superficially most successful form was therefore the culmination of evolution. But if Man is the best then why hasn't everything evolved into Man?

In the early nineteenth century Jean-Baptiste Lamark proposed that evolution progressed because offspring could inherit adaptations learned by their parents i.e. that the experiences of one generation were passed on to the next. Lamark was also big on progression – the idea that evolution was a directed towards ever more perfect forms, the ultimately perfect form of course being Man. These views have since been discredited and now Lamark is demonised as a bad scientist, but at the time his ideas were widely accepted and his observations on evolution had a great influence on what was to come. One feature of evolution which may give Lamark some late support is what you might call negative evolution e.g. the loss of flight in certain birds or the loss of eyes in some cave-dwelling or underground species. It is hard to see how losing some facility makes an individual more fit but if these facilities are lost simply because of lack of use, this smells of Lamarkism.

In 1858, the Englishman Charles Darwin published a book which shocked the world and has had a profound and lasting effect on the way we view evolution. The book was '*The Origin of Species*' and in it Darwin proposed the then radical view that life evolves by random mutation from pre-existing life forms and that these random mutations are 'selected' by nature – the 'unfit' being weeded out and the 'fit' surviving. Darwin applied this argument to the whole of the plant and animal kingdoms and included Man in it.

In almost all creation myths, including the Biblical one, Man is a separate creation with a special relationship with the gods. Darwin would appear at first glance to deny this special relationship and at the time of its publication '*The Origin of Species*' caused an outcry. There was lively public debate, the most famous debate being between Bishop Wilberforce on the creationist side and Thomas Huxley for the evolutionists. Of course Huxley won and the rest is history.

The argument of *The Origin of Species* in a nutshell is:

More individuals are born than can survive.
This leads to competition – the 'struggle for existence'.
All individuals vary and these variations are inheritable.
Some of these variations may make an animal more 'fit'.
The most 'fit' individual will survive and breed where the less fit will die.
As the most fit individuals will pass on the characteristics which made them fit, these characteristics will be propagated in the species.

As Darwin himself put it: 'As many more individuals of each species are born than can possibly survive; and as, consequently, there is a frequently recurring struggle for existence, it follows that any being, if it vary however slightly in any manner profitable to itself, will have a better chance of surviving, and thus be *naturally selected*' (Darwin's italics).

Darwin based his theory on the variations induced by selective breeding, on the success of feral species and observation of the world around him, specifically his observations while sailing around the world as the ship's naturalist and travelling companion of Captain Robert Fitzroy on the H.M.S. *Beagle*. He was fascinated by the intricacy of nature and the huge variety of creatures which inhabit every conceivable niche. He spent many years on *The Origin* and in fact only published the work when he heard that Alfred Wallace had come up with much the same idea and was about to publish himself.

Of course Darwin was a Victorian and his views to a great extent reflect the thinking of his times. Darwin was convinced in the doctrine of Uniformitarianism. This is the belief that the same forces which shape the world today – erosion, sedimentation, volcanism, earthquakes etc. – have always shaped the world and that, given the vast age of the Earth, all is explainable in terms of these forces alone.

The opposing view was Catastrophism which explained the world in terms of the biblical flood. Darwin was convinced that 'no cataclysm has desolated the whole world'. However, this debate is now open again and the catastrophists are on the up and up. We now know, or think we know, that sudden, catastrophic disaster has its place in the story of the world as well as gradual, incremental change. The demise of the dinosaurs is explained today by a massive meteor impact and disasters don't come much more sudden or catastrophic than that. There have been other waves of extinction and some scientists favour meteor impact as the vehicle for these as well. We may right now be in the middle of one of the biggest extinction events so far and it's not a meteor that's causing it.

Darwin's assumptions are firstly that there is a 'struggle for existence' in that '...every organic being is constantly endeavouring to increase in numbers...', and secondly that every individual varies to some degree from each other individual – one is bigger, hairier, brighter than the other. In most cases these variances are insignificant but in some cases they may give some advantage in 'the struggle for existence' in which case they will be perpetuated.

Much of Darwin's argument derives from observation of selective breeding by Man. Man has created woollier sheep, meatier cattle, fancier pigeons etc., by selecting individuals with the desired characteristics and cross-breeding them. We have, for example, produced breeds of dogs as distinct as Chihuahuas and Great

Danes. Of course these are all still the same species – a Chihuahua is still a dog, *Canis familiaris,* the same as a Great Dane is a dog and they could mate, albeit with difficulty, and produce viable off-spring. If they did breed then the off-spring would be neither pure Great Danes nor pure Chihuahuas but would have characteristics of both parents. The variations between the breeds have been induced only because the breeder has deliberately mated selected individuals.

In the natural world it may be that the biggest does not mate with the biggest and the smallest with the smallest so that the exaggerated results achieved by Man would be less likely to occur. Indeed, most feral animals quickly lose those Man-induced characteristics and revert to a generic archetype after only a few generations of inter-breeding with a wild population. Furthermore, in some cases the induced variations are positively inimical to the animal – the snub face of the Pekinese and bulldog make breathing more difficult for instance – and in other cases breeders have chosen actual mutants to propagate – hairless cats for example. These sorts of mutations would be unlikely to survive let alone thrive in the natural world.

Darwin is never clear exactly how these superficial differences may result in divergence into separate species. A species for the sake of this argument may be defined as a population that can only breed within itself. It is now taken that isolation is a major factor. If two breeding groups are split into separate populations, isolated from each other, then characteristics in each group may tend to become more exaggerated to a point where eventually the two groups can no longer inter-breed and produce viable off-spring. Evidence for this in the natural world comes mainly from islands where unique species have often developed. Since the advent of Man with his settlements and clearances there is obviously more chance of animal populations becoming isolated from each other. The world would naturally be a far freer and more mobile place if there were no roads or fences or canals or pipelines to impede progress. The argument is less convincing when applied generally. If we take, for example, the many varieties of woodland birds it is more difficult to see how isolation applies.

Darwin was also heavily influenced by the writings of Thomas Malthus whose book '*An Essay on the Principle of Population*' he had read in 1838. Indeed Darwin calls his theory 'the doctrine of Malthus applied to the whole animal and vegetable kingdoms'. In particular, Darwin's famous 'struggle for existence' is taken directly from Malthus although Malthus uses the phrase in reference to battles between human tribes rather than as a general description of the natural world: "...when they fell in with any tribes like their own, the contest was a struggle for existence, and they fought with a desperate courage..." Darwin imagined the struggle for existence on two levels: within a species there is a struggle between individuals for scant resources and between species there is an ongoing 'arms race' between predator and prey.

Whether Malthus and his ideas on human population pressure apply to the natural world is questionable. Animal populations are constrained in many cases by a limited breeding season and a great deal of natural wastage from starvation, disease, accident, predation etc. Animal populations, unlike human populations, do not tend to expand endlessly. Darwin's assertion that '...every organic being is constantly endeavouring to increase in numbers...' (cf. Malthus: '...plants and animals...are all impelled by a powerful instinct to the increase of their species...') does not really seem to be born out by the facts. Some animals, for example the kangaroo, can withhold from giving birth in times of drought. Others, such as lemmings, use mass suicide as a means of regulating population. In many animal populations, for example elephant seals or deer, only a very restricted number of individuals will breed at all contrary to Darwin's claim that 'amongst animals there are very few which do not annually pair'. Many animals and birds use territories to restrict numbers – only creatures that control a territory adequate to support a family will get the opportunity to mate. All of these strategies seem designed to limit numbers to a viable, supportable population rather than designed to 'increase in numbers'.

With the reduction in whale numbers over the last few hundred years and their current protection one might expect their population to be rapidly increasing but, while it is slowly creeping back from the extinction mark, it certainly isn't exploding as Darwin suggests when he says: 'Lighten any check, mitigate the destruction ever so little, and the number of the species will almost instantaneously increase to any amount'. The same could be said for bears, rhinoceros, American bison and dozens of other species. Elephant populations in some protected areas in Africa are certainly growing at fairly high rates – the population of the Kruger National Park in South Africa has doubled in recent years – but these are not natural populations in an unrestricted environment and it is impossible to say whether this rate of growth would be sustained in other circumstances.

The vast numbers of plankton in the seas, made up of embryonic animals and plants of all kinds could never all develop into adults – it is absurd to suggest that they are intended to. It is more plausible to regard them as fish food; that the very purpose of most spawn is not to grow to adulthood at all but to provide a feast for the many other animals that rely on them as part of the food chain. The world makes far more sense looked at in this way. There are several species, salmon for example, which die after spawning to become a feast for others. Nature is not a struggle, it is a bounty. Most animals do not actively struggle or compete for food; they passively browse on abundant resources.

We have been pillaging the seas for centuries, taking vast quantities of fish, yet although a few areas have become exhausted and some species are now becoming scarce, there are still fish to take. We have been felling forests for millennia but still there are forests to fell. The abundance of nature is not boundless but it is prolific. We take much, much more than our share of the world's output and yet the world still provides. If we were to only take our fair portion, there would be a huge surplus, enough for the whole natural world to survive without any struggle at all.

There is also the question of whether any particular doctrine should apply to 'the whole plant and animal kingdom'. Do the same rules which govern success and succession in the rapidly multiplying, asexual world of bacteria really apply to a blue whale or an orchid? Perhaps they do but perhaps that is because the mechanism of evolution is common to all living things and operates in a holistic manner: Darwin never offers any justification for his assumption.

Darwin bases much of his argument for the 'struggle for existence' on the runaway success of feral and introduced species where there are no natural predators, for example feral horses in America. The reason he cannot cite cases from nature and must resort to situations where Man has interfered is because his argument does not hold in the natural world. Indeed Darwin's argument that species are adapted to their environment might lead one to expect that naturally occurring species should be better adapted (fitter) than introduced species but this is exactly the opposite of what has actually happened – as Darwin himself points out.

There is a balance in nature between predator, prey and food source which tends to maintain each population in relation to the other. Ironically, the classic study of this comes from data supplied by fur trappers in nineteenth century Canada on hare and lynx numbers. As the lynx population increases and they kill more hares so hares becomes scarcer and lynxes begin to starve. As lynx numbers decrease so hares become more common and lynxes get an easier feed and their numbers increase, and so on. In nature the actual relationships are far more complex than this very simplified example. The whole ecosystem is one big feedback mechanism which tends to self-regulate.

There is a restraint in nature that is sadly lacking in Man. While some exceptional animals such as locusts will multiply to plague proportions and eat everything in their path, most animals do not outstrip the resources at their disposal nor do they destroy their environment and put pressure on themselves. The reputation of elephants as environmental vandals has been questioned by recent research which suggests that the destructive behaviour which has been observed only occurs because the animals have a confined range and cannot spread out as they would if unrestricted.

Even the so-called 'arms race' between predator and prey is questionable. Prey and predator for the most part live in an uneasy harmony. Most predators will target the old and sick or the young and the vulnerable in preference to healthy, mature animals that will put up strong resistance. They may be weeding out the 'unfit' but these animals are often literally 'unfit' in terms of health rather than 'unfit' in an evolutionary sense. In times of plenty everyone does well and when times are hard everyone suffers but, left to themselves, populations will stabilise.

Man has taken himself out of this feedback loop. He has few predators and those that still could kill him he keeps well at bay. If his food source runs out he grows more food. He has control of most of the diseases that attack him. He insulates himself from the vagaries of climate. If his teeth wear out or break or if they simply don't look nice enough, he can get false ones. He nurses his sick and wounded. He can mend bones, replace hearts, fix most things that would kill any other animal. He is fertile twelve months a year for thirty-odd years and can live to be a hundred. It is only Man whose population is out of control. Darwin's claim that 'Even slow-breeding man has doubled in twenty-five years' seems slightly ironic to the modern reader in the light of recent population statistics

The glib phrase 'survival of the fittest' also needs some explanation. The fittest does not necessarily mean the one that is strongest, biggest or fiercest; it means the one that fits best into the particular ecological niche it wishes to occupy. For example, the great white shark is superbly fit for the open ocean. It has many, replaceable, very sharp teeth in a jaw which it can dislocate to enable it to take the biggest possible bite; it can sense potential prey from a great distance; it has a streamlined body, fins, gills, a lateral line and all the other mechanisms which make it the supreme predator of the seas. However, put it in the middle of the Sahara desert and it would be dead in minutes. On the other hand, a camel is superbly suited to the heat and drought of the desert but in the deep ocean, being the only mammal which cannot swim, it would be nothing but shark food. The shark is neither fitter nor less fit than the camel – each has their niche and each is fit for it. They are both fit and both are survivors, indeed the shark has been around in much like its modern form for hundreds of millions of years. The reason both have survived is that they are well adapted ('fit') for their environment and lifestyle. It is meaningless even to talk of fitness outside the context of an ecological niche. We shall return to this point in our discussion of Man.

Darwin was unclear about what caused the variations between individuals. Since Darwin's time, the mechanism for these mutations has been recognised as the gene, and the mainspring of the gene is our friend DNA. Neo-Darwinism proposes that when a DNA molecule replicates (which they do a phenomenal number of times) that occasionally, just occasionally, it gets it wrong. Instead of making a perfect copy, there is a glitch, a mutation.

A DNA molecule is a miracle of information carrying. Imagine a blue-print for a human being and how many instructions that would need and then try to imagine this encoded in a twisted ladder structure in a sort of quaternary code. The DNA of, say, a flatworm is far, far simpler than that of Man. It is hard to explain how errors in copying could lead to an increase in information as seems to be the trend of evolution. Getting from flatworm to Man by mistake is a long shot at best.

There is also the unanswered question of what level or levels evolution may work on – individual, kin, population, species or whole world? The most popular choice seems to be the individual – that is certainly the way Darwin viewed it – but as family groups share the same (or much the same) DNA, some pundits favour evolution at this level. This makes altruistic behaviour within a kinship group fit better into the theory although it still leaves cooperative behaviour at a higher level out in the cold. Some propose evolution works at a species level but it seems that evolution makes most sense if considered at a whole world level. The whole world is inter-related. A change in one animal or plant affects not only that animal or plant but every other living thing that it interacts with. To try to explain the world by dissecting it and analysing the pieces risks missing the big picture.

Most neo-Darwinists take the two tenets of *The Origin* as gospel: that evolution is entirely random and that 'natural selection' is the only mechanism for change. Darwin was not so sure. He certainly held that evolution was random, that there was no 'arrow of evolution' but he did suspect that other unspecified forces may be at work. He took natural selection to be the main mechanism but in *The Origin* he says 'natural selection has been the most important, but not the exclusive, means of modification'.

To illustrate Darwin's theory, let us take the example of night vision in cats. We start with the assumption that cats were originally night-blind or at least did not have night vision as good as today. Then one day by pure random chance a mutant cat is born, a cat with eyes slightly more sensitive to low light than other cats. It can hunt later into the evening and thus can be more effective. It will therefore thrive and in thriving it will attract a mate (or many mates) and thus pass on its genetic mutation, its improved night vision, to its many progeny. The poor unimproved cats will be less effective and so will die sooner with fewer off-spring. This cycle continues. Eventually, over many generations, the cats with best night-vision will prevail and the night-blind cats will be weeded out. The 'fitter' cats will survive.

While effective night vision is obviously an advantage to a predator such as a cat, there are some developments which are less obviously advantageous. The classic example is the gaudy plumage of some birds, the ultimate instance of this being the peacock's tail. The peacock's tail is very beautiful but it is very large, which makes it cumbersome and very noticeable. This would appear to be a disadvantage in that a smaller, less obtrusive tail may make the peacock less easily spotted by predators and make it easier for the peacock to escape if being attacked. Why then has the peacock developed a tail which would appear to be positively detrimental to its survival?

Darwin's answer to this paradox was sexual selection. He proposed that peahens (which are, by the way, quite as dowdy as you might expect) have a fondness for big, bright tails and so tend to mate with the male with the biggest, brightest tail around. Why this should be so is less clear. Perhaps a big tail implies that the peacock is a vigorous, powerful bird who will be a good mate and father. Perhaps the peahen just likes the pattern.

While the peacock's tail is a plausible example of evolution by sexual selection, there are other examples which are less straightforward. Take the Australian bower bird which not only displays to the female but also builds a bower in which to display. Once he has built his bower, the male bird decorates it with coloured objects which he carefully selects and places, presumably in an effort to attract a mate. While it is quite reasonable to assume that the peacock's glorious tail is genetically inherited, it is harder to see how a bower bird's aesthetic taste could be so passed on. Or if aesthetic taste is just a matter of genetic inheritance then where does this leave Man's art and culture? Are these too simply sexually selected traits passed in the DNA from generation to generation?

There are many sexual displays and courtship behaviours in the animal kingdom which appear to have no other purpose than to attract a mate. Some animals, for example deer, have contests between the males for the sexual favours of the females. In many instances only the strongest (in this case this really does mean the fittest) will mate at all. Obviously an animal must mate to pass on its genes and thus sexual selection is a real and very important part of evolution, at least among animals where there is this sort of competition. This may perhaps explain some of the wilder excesses of evolution such as the Irish elk. This now extinct animal was blessed with a pair of antlers which would make any modern moose jealous. At up to 3.65m (12ft) from tip to tip and 40kg (88lb) in weight, the antlers reached an extreme of size, width and weight which must have had the poor males struggling to hold their heads up and the females fainting with excitement.

Sexual selection is certainly a factor in human reproduction and has probably been so from a very early stage of our development although the term 'sexual selection' is perhaps wrong when applied to Man – 'cultural selection' may be more accurate; in fact 'cultural selection' may be more accurate regardless of species. However, it is hard to be so sure about the sexual (or cultural) tastes of any other animal – I've never seen a particularly good-looking elephant although I'm sure they are out there. Sexual selection is almost always applied from a female perspective – it is males who have the bright feathers, the horns, antlers, tusks, manes etc. Most female displays are taken as advertising sexual receptivity rather than being competitive.

There is also a question as to why there need to be two selection processes – could we not manage with just one or the other? I see no reason why not. If a peahen can select a mate depending on his tail feathers then why cannot a giraffe choose a partner based on the length of its neck or a spider select a mate for its web-weaving expertise? There seems to be a redundancy here that Occam's razor could trim.

It seems that some traits such as mammals' warm blood and fur appear before their time is due. In the time of the dinosaurs, these features did not really give a great deal of advantage as the climate was mild but when the postulated meteor struck and the ensuing nuclear winter set in, these same features suddenly became life-saving assets and led to the rise of the mammals. Whether the mammals are therefore 'fitter' than the dinosaurs they supplanted or simply luckier depends on your viewpoint.

To answer some of the more difficult questions of evolution you need to go beyond the obvious and into the realm of speculation. Take, for example, the wings of insects. It is hard to see how a half-formed wing, too small and weak to allow flight, could give, say, a butterfly any evolutionary advantage. How then could the wing form by gradual steps if there is no advantage to be gained until it is fully developed? One answer to this conundrum is to suggest that the wing did not evolve for flight; it evolved for some other reason and then, later, became adapted for flight. A large surface with many blood vessels provides a fine cooling mechanism. This provides a plausible mechanism for the evolution of a large wing by gradual steps. Similar arguments have been made to answer many of the more awkward evolutionary questions.

The fossil record when Darwin was writing was extremely fragmentary. Darwin always assumed that the gaps in the fossil record would eventually be filled and that it would reveal the gradual gradation of intermediate forms that he predicted. This has not been the case. On the contrary, the fossil record seems to follow a general pattern of a long period of gradual evolution followed by a sudden extinction where a large percentage of all species disappear. After the extinction event there is a period of rapid evolution where many new species suddenly appear. This is followed by another long period of gradual evolution and so on. This has led to a variation on the Darwinian theme known as Punctuated Equilibrium. This theory accepts the two vital precepts of Darwinian evolution – random mutation and natural selection – but instead of the gradual, cumulative changes conceived by Darwin, suggests that evolution goes in fits and starts. The mechanisms which regulate these sudden bursts of evolution are unclear but this does seem to fit with the fossil record slightly better than Darwinian gradualism. Darwin specifically disagreed with this view: '[nature] can never take a great and sudden leap, but must advance by short and sure, though slow steps'.

It is ironic that Darwin's theory should still be regarded today as the best available when several of the ideas on which the theory was based are questioned. The fossil record, far from providing a picture of gradual change with few gaps between forms, seems to show progress by leaps and bounds. Prehistory seems to be punctuated with cataclysmic catastrophes. Darwin's theory was based to a great extent on theories about human population pressures, on observations of human selective breeding and on the success of species transplanted by humans to alien environments. All of these are questionable when applied to the 'natural' world. Darwin did not know the mechanism for his mutations although the subsequent discovery of genetics and DNA does provide this mechanism.

One criterion for a scientific theory to be acceptable is that it should be possible to prove or disprove. A major problem with Darwin's theory is that it is impossible to prove although it is equally impossible to disprove. We simply do not live long enough to see evolution at work. There is a claimed example of evolution in the change in colour of pepper moths in the UK since the industrial revolution. The story goes that the moths became darker as the sooty waste from industry darkened the surrounding environment and have since become lighter again as the Clean Air Act has worked its magic.

However this is only half the story of evolution: it may well demonstrate survival of the fittest but it does not demonstrate random mutation. The distribution of lighter or darker pigments in the population of moths forms a normal bell-curve i.e. the majority of moths are coloured somewhere in the mid-range with fewer extremely light or extremely dark moths. All the above story demonstrates is that the bell-curve of distribution moved slightly left (lighter) or right (darker). There were always darker moths and lighter moths but these moths are genetically very similar or at least only minimally variant. The lighter or darker moths may predominate due to natural selection but we have not seen any random mutation. Certainly the darker moths do not represent a new species of moth. Remember that evolution is not just about a particular species of animal developing darker colouring or longer legs, it is about species giving rise over time to entirely new species: a proto-type cat becoming a lion and a tiger and a leopard or a dinosaur becoming an ostrich and a parrot and a humming bird.

Returning to our night-vision cat, there are other objections to the scenario. The most obvious is: is it likely that effective night-vision will spontaneously develop from a random mutation? The answer to this is of course no. But the saving grace of evolution is time. Given a huge number of cats and a huge amount of time, almost anything is possible, even something as unlikely as night-vision spontaneously developing. Whether the numbers and time involved are large enough to make this scenario credible depends again on your point of view. I have read convincing arguments both ways.

The cat also has to realise it has good night vision and change its hunting behaviour to take advantage of this vision. This is a factor in many evolutionary developments, for example it is all very well for a spider to be able to produce silk but it must then learn to spin a web. Behavioural changes may be more difficult to explain than physical changes. It is relatively easy to see how a change in the synthesis of certain proteins could affect the physical characteristics of an animal and to believe that these changes could be carried on to its progeny. How similar physical changes could result in changes to behaviour, how these behavioural patterns are passed on and, most remarkably, how the physical and behavioural changes are coordinated is less clear.

And then there is the issue of pure, dumb luck. Most animals do not survive to adulthood through accident or illness, starvation or predation. Fish, for example, may produce thousands of eggs but most of these will die before they hatch and of the ones that do survive only a handful will finally make it to adulthood. A mutant with an improvement to, say, the fins of the adult fish would have to make it through the lottery of the early years before its benefit could be felt. Our special cat would need a good deal of good fortune to survive. As for the poor peacock, it is only when it reaches maturity that its splendour will be seen. Until then it takes its chance with its less flamboyant brethren.

There is also the question of how much benefit does the cat gain and would this be significant. Obviously its immediate forebears survived perfectly well before they had night vision. A cat only needs so much to eat and if it can get this without night vision then will night vision make it significantly 'fitter'? These arguments apply to almost all evolutionary scenarios. Other more serious objections apply to some.

Take, for example, Darwin's famous finches. One of the ports of call of Darwin on his voyage of discovery on the *Beagle* was the Galapagos Islands. On each island were birds which were obviously (to Darwin) related but which had different beaks, claws and feeding behaviours. Darwin conjectured that these were all derived from a single original form but they had specialised in different areas and evolution had adapted each one to fit snugly into their own niche. The thing is that mutating a beak to crack a nut instead of probing for nectar is fine but you also need to change the digestive system to eat nuts and the behaviour to feed on nuts. A bird which can eat nuts but not digest them or which tries probing flowers with its nut-cracker beak will not do well.

Although we may understand that such and such a gene produces such and such an effect, we have very limited understanding of the actual mechanisms involved. We know that sometimes a change in one gene can cause another to behave slightly differently or not to work at all. It is quite possible that changes to beak-shape, digestive system and behaviour could all be controlled by the same gene complex and changes could be a package deal.

There are many examples of these sorts of 'packaged' adaptations and some extend inter-species. For example, flowers produce pollen to attract insects which spread the pollen and pollinate the flowers. In some species, for example the bee orchid, this is taken to extremes. The flower of the bee orchid resembles a bee. A bee sees the flower, tries to mate with it and gets covered in pollen. The next flower it tries to mate with receives not only its unwanted affection but also its load of pollen. This requires a good deal of synchronised evolution between the flower and the insect. There are even more extreme examples of co-evolution in some fig trees which are exclusively pollinated by species of wasps which live nowhere else but in these trees and whose whole life-cycle from the moment they hatch to the time they lay their own eggs is intimately and inextricably intertwined with the tree.

There is also the question of why the particular variations which exist today have developed and survived and others have not. To return to Darwin's finches, it is highly unlikely that every possible genetic variation on the finch theme has actually been tested in the evolutionary laboratory. Some variations have happened and some have not – there are no aquatic or flightless finches for example. Are these variations not possible or not viable? Are the varieties that we have today really 'fittest' or are they simply a random selection of an infinite number of possibilities? Or are they transient stages in the evolution of new varieties of finches yet to come?

Of course, while random mutation combined with selection could produce the results we see, so could non-random mutation. The fact that evolution could be random does not imply that it necessarily is. In any theory, the simplest answer which fits all the facts is taken to be the best, but it is questionable whether random mutation and selection (plus sexual selection, pre-adaptation and all the other supplementary bits and pieces needed to make neo-Darwinism work properly) is 'simpler' than postulating some directing force. By 'directing force' I am not implying Intelligent Design. A salt crystal for example is certainly not random but neither is it 'intelligently designed'. Its form is constrained by the laws of physics and chemistry and the inherent structure of the crystal. I can see no reason why mutations in DNA may not be constrained by some similar, as yet undiscovered deep structure.

One of the most striking features of Darwin's theory is the fact that it is supported through thick and thin. No matter what evidence is unearthed and no matter how that sits with *The Origin*, it is absorbed into the theory. Darwin is an icon who cannot be questioned. Even if he was wrong on catastrophes, even if he was wrong on the completion of the fossil record, even if he was wrong on evolution by leaps and bounds, no matter what, he is supported. His name is invoked in the most inapplicable circumstances to legitimise any theory. Society and culture do not mutate at random, they do not pass on characteristics to their progeny, they are not in any 'struggle for existence' and yet we have Social Darwinism. People talk of the 'evolution' of, say, the motor car as if the latest models have been naturally selected.

I believe this is for several reasons:

Firstly Darwin's theory is so beautifully self-contained: nature is as it is because it selects itself – no gods, no directing mechanism, nothing but nature itself.

Secondly, Darwin can be used to explain anything: If a species produces many young then this aids survival by providing redundancy; if a species produces few young then this aids survival because the female can concentrate her effort on her fewer offspring. If a snake slithers or a human walks or a whale swims or an ape brachiates then this aids its survival in one way or another. Any trait in any animal can be justified by survival of the fittest as it stands to reason that if a trait exists and the animal survives then the trait must assist survival. Even such seeming anomalies as the strap-tooth whale are explained. The strap-tooth whale has tusks but these tusks in adult males grow over the top of the snout preventing it from opening. It is hard to see how this assists survival but the whale survives (at the moment) so it must be a good thing.

The assumption is that all animals living today are survivors, but this is not necessarily true. It is true that they survive today but what of the long term? Perhaps the strap-tooth whale is an evolutionary dead-end which is just taking time to be eliminated in the 'struggle for existence'. It is as if we are looking at a few frames of a film and trying to guess the story from what we see. As I said at the beginning: now is not the end of the story, it is simply an arbitrary point along the way.

Thirdly, Darwin still leaves Man at the top of the evolutionary tree. Admittedly there is no real reason why we should regard Man, specifically the particular instance of Man which happens to be extant at this particular moment in time, to be the pinnacle of progress but even Darwin says in *The Origin* '…what is meant by an advance in organisation. Amongst the vertebrata the degree of intellect and approach in structure to man clearly come into play'. Man is still made in God's image. Being the dominant species he is seen as the winner of the evolutionary race. Who is running second is not so clear.

Fourthly there is obviously a lot of truth in the theory. Evolution is a proven biological fact no matter what the Creationist lobby may claim. Animals and plants do vary and characteristics are inherited by their offspring. An animal born with a debilitating genetic mutation will probably die so it is true that genetic weaklings are weeded out although the stronger claim that a genetic improvement will necessarily be preserved is not proven beyond doubt.

And lastly, nobody has come up with anything much better.

The most serious problem is that if you reject random mutation and natural selections as mechanisms for evolution then you must propose some other mechanism. The main contender is God or some other directing force (Intelligent Design) but this tends to lead to more and more issues. The very mention of a directing force is heresy in the scientific world and even worse is Creationism – the belief that a god or gods created the world. This is usually treated with scorn and derision and rarely with serious debate. Despite the anomalies in, objections to and lack of evidence for neo-Darwinism it is more acceptable to the scientific community to keep to this faith rather than even discuss the 'c' word. And at the end of the day it is a matter of faith.

One alternative view is to regard the world as a single entity, the so-called Gaea thesis. Instead of regarding life as a struggle for existence, we can regard it as a cooperative venture. There is no reason to believe with Darwin that all animals are competitive and that no animal does anything for other than entirely selfish reasons. Why could a fish not over-produce in order to provide a meal for other fish? This is entirely as plausible as suggesting that '...every organic being is constantly endeavouring to increase in numbers…' and that '… [it] increases at so high a rate that, if not destroyed, the Earth would soon be covered by the progeny of a single pair'. There is absolutely no evidence to support this view.

In the original Gaea thesis, the Earth was regarded as an actual single living entity. This interpretation has been largely discredited but the weaker interpretation that all life is a single, interrelated network and evolves in co-ordination instead of competition, is quite tenable. The oxygen producing prokaryotes heralded the opening of the world to other life. The burgeoning of life itself creates a new environment for new life which creates new environments for new life and so on. The co-evolution of insects and flowers is natural; the balance of predator and prey is to be expected. The fact that all living entities share a common genetic heritage means that all living things are related and this relationship applies on more than a superficial level, it may itself constrain evolution. The feedback loop of nature may work more on what is best for the whole rather than what is best for one selfish individual.

Or perhaps, if the whole of nature is one big feedback loop then could the result be chaos? I do not know if chaos theory has yet been applied to evolution but this seems entirely as plausible an approach as any other. The world is full of coincidences and happenstance and interactions and unlikely events. Punctuated equilibrium may fit better with the peaks and troughs of chaotic progress than with other more gradual models.

My main objection to Darwin is not to his theory; it is to his portrayal of the world as a struggle for existence. This view of 'Nature, red in tooth and claw' is used to justify our own rapacity and ruthlessness. It is a uniquely human perspective based on human ego, human self-awareness and human selfishness. It is a view of the Victorian big game hunter and its day has long passed. Really this view is not justified. Our whole attitude to nature has now changed. The more we learn about nature, the more intricate and marvellous it appears. Examples of co-operation are as common as examples of confrontation. In Darwin's time, some knowledge of the natural world came from dedicated study such as Darwin's own experiences on the *Beagle,* but just as much came from received wisdom and apocryphal travellers' tales. The standard, depth and breadth of nature study today are hugely greater than anything Darwin had access to. Surely it is time to review some of his assumptions in the light of this knowledge.

Evolution is a proven fact: most living things have become more sophisticated over time. However I do not believe that this is due to competition and struggle. Life is not a never-ending battle for survival between individuals. If a lion kills a baby antelope this can be viewed not as a 'struggle for survival' but as nature feeding nature: it is just as plausible to see this as cooperation between the antelope and the lion as to see it as a battle. Furthermore any theory of natural evolution should ignore Man altogether. Man is so far outside the mainstream of evolution and has developed so rapidly and so radically that it is hard to fit him into a general schema. And the fact that at this stage he is still a newcomer to the world rather than a proven survivor means that there is no need to incorporate him into a theory of long-term development.

How life began and whether all living things have a family tree going right back to the first protozoan are matters on which I keep an open mind. But I believe, as did Darwin, that there are more forces at work than plain, dumb, random mutation and natural selection. Our understanding of evolution is at best incomplete.

The truth of the matter will always be up for debate and perhaps in future an evolutionary theory will be adopted which supplants Darwin. However, for the moment he is the best we have and we will work with his theory.

A theory should provide a basis for prediction. Darwin's theory does not on the face of it predict much more than that evolution will continue but if we accept Darwin then there are four clear implications of the theory which may provide some basis for prediction. These implications are largely overlooked or ignored.

Life develops by random mutation. If we accept this then we must accept that there is a far greater chance of an inimical mutation than of a beneficial one. Most common genetic mutations result in defects such as spina bifida or mongolism. It is far easier to imagine a mutation that would for example impair vision than one that would improve it. And with any genetic mutation there may need to be behavioural changes without which the mutation is useless. Thus, at any given point in time, the theory would predict that we should expect to see more negative mutations than positive. To return for a second to the Irish elk: the exaggerated headgear of the elk has been cited as an example of sexual selection but perhaps it is not. Perhaps it is simply a mistake, one of the many negative mutations that must happen all the time. The elk developed its antlers over a few hundred thousand years. It did not last a long time in the grand geological timescale of evolution. It could simply be a short-term success but a long-term failure.

Fit means fit for a specific ecological niche. One of the main sources for Darwin's work was the variety of finches in the Galapagos Islands. These were all derived from a single original form but they had specialised in different areas and evolution had adapted each one to fit snugly into their niche. No one finch is 'the fittest'. No finch is more or less fit than any other; each simply fits in a different place. The implication is that fitness is only meaningful in the context of a niche; an animal that does not have a niche cannot be fit.

It is survival of the fittest not survival of the fit. The fittest is the best of those competing for a niche. If there is little or no competition, there is little struggle for survival and so evolution may let some things go which may be punished under other circumstances. Take for example the birds of New Zealand. In New Zealand there are few native land mammals and so birds have taken over many of those niches normally occupied by smaller animals. An example is the kakapo, a ground-dwelling vegetarian parrot which in many countries would compete with, and lose out to, the rabbit. The kakapo evolved from a parrot which could fly. However flying is expensive in terms of energy and over time, with no need to fly, no apparent evolutionary advantage to be gained, it has lost the habit and the ability. The inability to fly does not impinge on the kakapo's chosen lifestyle and so, although the loss of flight can hardly be called an evolutionary advantage, it is still the fittest competing for its niche. That was until the arrival of Man with his accompanying vermin. The flightless kakapo is now on the verge of extinction because the hunters of rabbits – cats, foxes, weasels, rats and Man – have invaded its comfortable island sanctuary. The loss of flight now doesn't seem like such a good idea. A flying kakapo would now swamp its flightless brothers. Thus the theory would predict that you're only as good as the opposition and that in the absence of opposition, you may not be as good as you think.

Evolution takes place over a very long time scale. The only reason Darwin could propose random mutation and 'survival of the fittest' as mechanisms of evolution is because there was time for it to have happened. With a world four thousand years old, there is no time for things to have changed radically. With a world several billion years old, anything is possible. Again and again Darwin emphasises the vast ages required for his gradual build up of beneficial mutations. With the conjectural evidence for evolution by 'Punctuated Equilibrium' the time scales may be shorter but even the rise of Man has taken a few million years and that has been sudden compared to the evolution of most species. Whatever the truth of the matter, it is a fundamental of evolution that it is a long-term process measured in millions of years rather than hundreds. Thus the theory predicts that short-term success is no success at all.

We will return to these points in our consideration of human evolution but now, back to the story…

When we last looked at life, it had got as far as the flatworm and its buddies. Now the pace of the story hots up. Only about fifty million years after the Cambrian Explosion we get the first primitive fish. These are also in our line of descent because they are the first animals with backbones (vertebrates). They dominated the oceans for millions of years and there were plenty of oceans for them to dominate. The world at this time was in fact mostly ocean as all the land was massed in one great lump called Pangaea. Over time this super-continent broke into bits and these pieces started drifting about and crashing into each other for the next few hundreds of millions of years to give us the world we know today. The same forces which broke up Pangaea are still at work today. Continents are still splitting, drifting and crashing albeit very, very slowly – only at about the same speed as your fingernails are growing. In another few hundred million years the world will be a very different place.

After fish we start getting amphibians and then reptiles and insects. Plants meanwhile developed from simple algae into ferns and forests and took over the land. It is the remains of these massive forests that we are now burning so recklessly as coal in our power stations.

Life on Earth has gone through periods of great innovation and episodes of massive extinction. There have been (so far) five massive extinctions where a significant number of all species on Earth have disappeared and many other less massive events when a smaller, but still significant number of species have gone. It has been suggested that these extinction events occur at regular intervals of about 26,000,000 years and that we may be due for another one about now. We may even be in the middle of it. We may even be causing it.

The largest single extinction event the world has yet known, the cataclysmic Permian extinction in which between 70% and 95% of all species disappeared from the fossil record, happened about 250,000,000 years ago. It was after this event that the first dinosaurs arose and they dominated the globe, both on land and in the seas, for the next 180,000,000 years. Flowers appeared on the land and insects boomed. The first birds appeared. The first mammals began to creep about beneath the dinosaurs' mighty feet to find themselves humble graves. And then, about 65,000,000 years ago the dinosaurs disappeared completely.

The Cretaceous extinction is the second largest of all extinctions (so far). In the course of a few thousands, perhaps a few millions of years, some 65% of all species became extinct. Among the victims were the dinosaurs. This catastrophe is marked in the geological record by a thin stratum of soil rich in iridium, the famous K-T boundary (KT stands for Cretaceous-Tertiary in German). Iridium is a mineral, rare on Earth nowadays but relatively common in asteroids.

There have been many theories put forward to account for this sudden disappearance of species which had been around for the last couple of hundred million years – climate change, disease, volcanism – but the current favourite is a meteor impact. About 65,000,000 years ago, off the coast of the Yucatan peninsular in Mexico, a comet the size of a small island hit the Earth. The debris from this explosion blackened the sky and poisoned the Earth in a nuclear winter of unimaginable proportion and blanketed the world with a thin layer of iridium. The small-bodied, warm-blooded mammals survived this cold, bleak period but for the larger-bodied dinosaurs it was curtains. The forests they ate died. Their huge bodies suffered from the cold. In a few brief million years, they were gone.

I should warn you that this is only the current theory. When I was a boy, dinosaurs were solitary, lumbering, cold-blooded, dumb reptiles that died out, probably because of climate change. Now dinosaurs are herd-dwelling, quick-moving, perhaps warm-blooded, intelligent proto-birds who were suddenly wiped out by an asteroid collision. Perhaps the truth lies somewhere in the middle. The quick, intelligent lobby is certainly more popular nowadays – quick, intelligent dinosaurs look so much better on film than slow, dumb ones and meteor impact makes much better television than disease or climate change. Quite probably it was a combination of factors which killed off the dinosaurs, but perhaps there is a different answer.

I have a theory:

Dinosaurs originated somewhere back at the beginning of the Triassic, about 250,000,000 years ago. Over the next 180,000,000 years or so they evolved and flourished to become the dominant species. Dinosaurs ranged from Compsognathus, as big as a hen, to Gigantosaurus (or Megalosaurus or Collososaurus or whatever is the biggest today), as big as a four-storey building, with the full gamut in between. There were plant eaters and eaters of plant-eaters and eaters of eaters of plant-eaters.

Of course, in their long, long period of dominance on Earth many dinosaur species came and went. The less fit perished and the best and brightest survived. And at the end of the Cretaceous, the cream of the crop, the boss hunters, were the raptors. There were various species but for this story we will concentrate on one: Deinonychus – those rather violent creatures in *Jurassic Park,* with a hooked claw on the hind foot.

Deinonychus was a raptor of the late Cretaceous. As depicted in the film, it walked upright and had good binocular vision. It hunted co-operatively. It showed intelligence. It may have been warm-blooded or at least not entirely cold-blooded. It had a useable hand.

Suppose that in a micro-burst of evolution, a mere one or two hundred thousand years say, a new species, *Deinonychus arrogans*, were to develop. Suppose this super-raptor were to develop intelligence and then language and began to change its whole way of life. Suppose they formed into clans, communities, eventually into organised cities. Suppose they domesticated the brontosaurs. Suppose numbers grew. Suppose they continued to develop and expand and eventually to conflict. Suppose that countries were invented and wars were declared.

And suppose iridium was not as rare back then as it is today. Suppose there were deposits of iridium which could be mined, and that they were mined. Suppose weapons were made, dreadful, doomsday weapons of such power that they would eradicate the world if they were released – iridium bombs. Suppose they were released.

Voila! The K-T boundary and the extinction of the dinosaurs.

Unfortunately as all traces of *Deinonychus arrogans* would have been totally obliterated by the release of these terrible weapons, my theory can never be proved (nor conclusively disproved).

Of course this theory is absurd. No species could change so radically from being just another animal to totally dominating the globe in just a few short hundreds of thousands of years. No species could be so clever as to develop technology capable of destroying the world and at the same time be so stupid as to use it.

Or could it?

Read on…

Man versus Comet

Man has the destructive power at his command to destroy this planet. He has stockpiles of nuclear, chemical and biological weapons enough to annihilate us all. He has his finger on a hair trigger.

The destruction of the massive US and USSR stockpiles have not made the world a safer place. While there were enough weapons to ensure that any aggressor would be properly punished by a retaliatory strike (Mutually Assured Destruction) then all we had to worry about was a madman or a mistake. Now with fewer weapons in more hands the thought of a pre-emptive nuclear strike is not unthinkable. The most likely scenario would be a nuclear strike at a nuclear target. This is less destructive than a nuclear strike on a civilian target but it is still not something most people would want to see. The danger then would be that once one nuke has been fired then the next one is much easier and then easier still and then hard to stop. If the unthinkable were to happen and the ball went up and every man and his dog with a nuke in his pocket threw it into the global melting pot and if it really got dirty and biological and chemical were used then...

Then even if the event went global there would be survivors – the few humans still living in the isolated places of the Earth plus those fortunate enough, or rich enough, to find shelter. They would crawl back into a world of radiation, disease and poison where the infrastructure of Man had been destroyed. If there was ever a rich opportunity for evolution then this surely would be it. Who would be the winner in the next lap of the evolutionary race – *Homo tristis sapientor* (sadder and wiser Man) or *Blattaria irradiata* (nuclear cockroach)?

Compare this to a comet impact. Suppose something the size of Texas hit the Earth. Chances are it would hit an ocean – they do cover more than half the world. This would create tidal waves which would wipe out most coastal settlements and seismic overtones which would rock the world. Debris would blacken the sky for months, perhaps years. Crops would die. Billions of people would probably die in the first few minutes and billions more over the course of the next months and years. There would probably be a few survivors – those hardy enough and ruthless enough to hang on through the long nuclear winter. Perhaps the creatures of the night would become the new lords of the Earth – or perhaps *Homo sapiens superstes* (surviving wise Man) would prevail.

There are contingency plans to try to avert any such impact. Telescopes scan the sky searching for stray rocks heading our way. Whether we could effectively destroy or divert a rock the size of Texas is a point that is hotly debated. Whether we should or not is rarely questioned but perhaps it should be. We are after all the ultimate winner from the comet impact which finished off the dinosaurs.

A Skew in Evolution

Is it a pithecene? Is it a hominid? Is it Neanderthal?
Is it Man Who Knows?

Before discussing the emergence of Man may be best to define what we mean by 'Man'.

The current scientific designation of modern Man is *Homo sapiens*. This is usually translated as 'Man who knows' or 'wise Man'. Man belongs to the genus *Homo*, which is part of the family of hominids which is part of the order of primates which is part of the class of mammals which is part of the phylum of chordates which is part of the animal kingdom. We are the only species of hominid now living on Earth, the last twig on the hominid tree.

Up until a few years ago, modern Man was classified as *Homo sapiens sapiens,* translatable, roughly, as 'Man who knows and knows that he knows'. Neanderthals were known as *Homo sapiens neanderthalensis* i.e. they were regarded as a sub-species of *Homo sapiens* on the same level as *sapiens sapiens.* Older *sapiens* fossils were classified as archaic *Homo sapiens* and this was assumed to be the parent species of both *sapiens sapiens* and *sapiens neanderthalensis*. Genetic analysis has led to Neanderthals now being regarded as a separate species and the distinction between archaic *sapiens* and *sapiens sapiens* being dropped.

All humans belong to this same species be they Australian Aborigine, Eskimo or English; whether they are black, white or yellow; whether they live in a mud hut or a penthouse. Humans as a species are remarkably homogeneous, much more so than most other animals. All humans share over 99% of their DNA. There is less genetic variation in the whole human race than in an average herd of gnus wandering the African plains. This is not really so surprising when you consider the relative recent origin of our species, as there simply has not been sufficient time for substantial genetic change.

There are various races with some noticeable differences between them and much has been made of these differences but the difference between, say, a Greenland Eskimo and a Papua New Guinea highlander are as superficial as the differences between a German Shepherd and a Pekinese. Of course, to a dog breeder the difference between a German Shepherd and a Pekinese is significant – pedigrees matter and pure blood lines are of paramount importance: a cross-bred dog is a mongrel and is less valued than its pure-bred cousin. Similar views of the human race are held by some people who like to imagine one race (usually their own) as being purer than or superior to others. There is no scientific justification for such a view.

It is possible that racial differences may be a precursor to species evolution. Suppose the Eskimos or Australian Aborigines or African Pigmies had remained isolated for a few more hundreds of thousands of years then perhaps we would have seen more species of humans arise. But this did not happen and is unlikely to happen in the near future as the world gets smaller and smaller and there is ever more interaction between all people.

In the far distant future the question is more open. Say a spaceship full of hopeful settlers is sent from the Earth to colonise some distant planet. Isolated and alone the settlers live, breed, die and evolve on their own world. What form of Man would be there after a million years? Would they last a million years? Will we?

Before trying to give even an overview of human evolution, I must warn you that the human family tree is involved and disputed. There are many unanswered questions and awkward gaps in the record. Our information comes from a relatively small number of fossils – a few thousand pieces coming from a few hundred individuals at most – and most of these fossils have been found at only a handful of sites. Most fossils are fragmentary and scattered; complete skeletons, especially of our early ancestors, are rare. Classification of fossil finds is often a matter of opinion or conjecture. The numbers of species, the relationships between them, where they originated, how and when they dispersed, the timing of the development of language, where the Neanderthals fit into the picture etc., are all questionable. A further level of complication is provided by genetic evidence which often corroborates the fossil record but which is sometimes at odds with it or is open to other interpretation. Just about every claim made about human evolution is disputed.

Palaeoanthropology is a very emotional subject. People get very excited about the deep origins of Man and how he fits into the grand scheme of things. Some people want to see Man as a natural part of the skein of life, nothing more than a sophisticated animal; some want to see him as a child of the gods, separate from nature and the master of all; some want to see him as the pinnacle of evolution, the culmination of an inexorable progression towards perfection. Personally, I see Man as an aberration, an outsider, a knot in the fabric of life. These viewpoints have as much or more to do with philosophy as with science.

Palaeoanthropology is also ever-changing. In the last few years there have been several major finds: *Orrorin tugenensis*, so-called Millennium Man, found in Kenya in 2001 and *Homo floresiensis*, so-called Hobbit found in Indonesia in 2003. The first of these is

65

significant in that at about six million years old it is potentially the earliest ancestor of the hominid line, pre-dating the australopithicenes. The second find is significant in that it indicates there may have been another dwarf species of *Homo* living alongside us as recently as 18,000 years ago. Very recently claimed 'human' remains have been found in Israel dated at 400,000 year old. This is long prior to the previous oldest finds and long prior to the supposed exodus from our supposed African origins. If proved correct, this may change much of the story I am about to tell but, of course, all of these claims are disputed.

The version of human evolution given here is simplified in the extreme. Without wishing to sound like a legal document, no firm relationships are stated or implied; all dates should be viewed with extreme caution.

The archaeological evidence of Man's development is ambiguous and far from complete. There are gaps of millions of years in the fossil record at the critical periods of Man's divergence from other primates and it seems that often as these gaps are filled, the picture becomes more complex rather than simpler.

To go along with the archaeological finds, our knowledge of DNA is growing at an enormous pace. The human genome has been decoded to a greater or lesser extent. Comparative studies of human and ape DNA have revealed new evidence concerning our relationship to other primates. Some DNA fragments extracted from Neanderthal fossils have been sequenced and compared to modern human and chimpanzee DNA to try to sort out the continuing enigma of their place in the scheme of human evolution. Other research is being carried out into mapping the distribution of the human race by DNA.

As most evidence of human evolution comes from fossils, it seems only reasonable that before getting down to cases we should first briefly digress into what fossils are and how they happen. Fossils are the traces in the geological record of animals and plants. They can be as complete and unequivocal as a full sabre-tooth cat skeleton preserved in the La Brea tar pits or as ephemeral as the trace of a worm's burrow preserved in sedimentary rock.

Fossils are only formed in special circumstances where conditions are favourable. When an animal or plant dies, under normal circumstances its body will either be eaten by scavengers or will decay and eventually will simply rot away. However under some circumstances, if for example it is covered soon after it dies in a layer of mud which protects it from other animals and from the elements, then it may be preserved as a fossil. Even if an animal does leave fossil remains the probability is that the fossil will be incomplete – complete skeletons of any fossil animal are extremely rare. What is more, only hard parts of a body will fossilise. Soft tissue such as muscles, internal organs, brains or tongues are very unlikely to leave any trace at all. Then once the fossil is formed it is subjected to all the vagaries of the Earth – erosion, upheaval, earthquakes, and volcanism.

Once a fossil is established we have the second half of the problem – finding it. Fossils become exposed through erosion or upheaval or excavation or a combination of these elements. Which fossils become exposed, upheaved or excavated is largely a matter of chance. Fossils may be destroyed during excavations for building or road-making or mining. It is only in the last couple of hundred years that fossils have even been recognised as something significant. Prior to this time and probably many times since fossils have been dug up and then thrown away as being nothing more than old rubbish. In modern times, even if fossils are actually exposed and recognised as fossils it is still not certain that someone will abort what might be a multi-million-dollar development so that a site can be designated 'of archaeological significance'.

Palaeontologists tend to look for fossils in places where the conditions favour fossilisation and where the conditions favour finding fossils. This is a tiny percentage of the world. For example, there have been many fossil finds in east Africa around the Rift Valley where the arid climate helps preservation and movements in the Earth's crust have exposed many old rocks. Conversely there have been few finds in tropical Asia where high humidity lowers the chances of fossils forming and dense jungle hampers exploration.

Having found a fossil, now come the real problems – identifying it, classifying it, dating it and, biggest problem of all, fitting it into the mosaic of current information in its proper place. Remember that many fossils are tiny and scattered – a fragment of a skull here, a bone over there – Java Man for example was first identified by only a skull cap and a femur. The most complete early hominid fossil ever found – 'Lucy' from east Africa – is only 40% complete.

With more scientists looking at more fossils with better techniques, the fossil record is slowly being filled in but it is still far from complete. There are millions of years of evolution which are simply unaccounted for because no fossils have been found.

Personally I am amazed not only that fossils of animals millions of years old exist but at how good they are and how much can be inferred from tiny fragments of bone.

But to continue the story…

The demise of the dinosaurs left the field open for a new group to rise. The group who grabbed this opportunity with all four paws was the mammals. Mammals had existed for some time before the dinosaur extinction but only as small, bit-players in the grand opera of life. When nuclear winter came in the wake of the Yucatan meteor impact the mammals with their fur, their warm blood and their quick brains were perfectly placed (pre-adapted?) to grab the opportunity. Over the next few million years there was a burst of evolutionary diversification as mammals took over many of the niches vacated by the dinosaurs.

One of the most successful mammal families was the primates. The early primates were small shrew-like creatures which lived in the trees. They had many of the characteristics of modern primates – grasping hands, wide-spaced, forward-looking eyes implying they had well-developed binocular vision and, most important from Man's view, the beginnings of our swollen brain. Over the next 50,000,000 years primates evolved in several directions into lemurs and lorises, tarsiers, monkeys and apes.

The ones we are most interested in lived about 20,000,000 years ago and belonged to the genus *Driopithicus*. These were ape-like creatures that lived in the rainforests which then covered much of Africa. As the climate cooled and the forests gave way to more open woodland and open savannah, so a new ape genus gained precedence, *Ramapithecus*. There were lots of these apes around between about 16,000,000 years ago and 8,000,000 years ago and they were widely spread. Then there is one of those annoying gaps in the fossil record.

Apes and humans are undoubtedly related but exactly when hominids (humans) and pongids (apes) diverged is still open to question. The search for a 'missing link' has been going on since Darwin first suggested that apes and humans were related but this is really a chimera. Since the split, which has been postulated at anywhere between twenty odd million years ago and as little as four million years ago, apes and humans have diverged more and more. The 'missing link' is an ancestral form which preceded the split. There have been many pretenders but so far there is no single species which has certainly claimed the crown. Some of the contenders are *Aegyptopithecus* who lived about 28,000,000 or so years ago, several species of *Proconsul* and an animal called *Kenyapithecus wickeri*. There are probably others.

Before going on, I should point out that there is a problem in deciding just where to draw the line between ape and human.

Man's closest relatives are the great apes, and closest of the great apes are the chimpanzee and bonobo (formerly known as the pygmy chimpanzee). We are surprisingly closely related to chimpanzees – we differ in less than 3% of our DNA – albeit that 3% of a genome as complex as the human genome is still a significant amount. Perhaps even more surprising is that chimpanzees and bonobos are genetically more closely related to humans than they are to other apes, even to gorillas. What this means is that gorillas split away from the human/chimp evolutionary line before chimps and humans split from each other. Of course, our understanding of DNA is still in its early stages and the fact that there is less than a 3% variance identified between Man and chimp may lead less to the conclusion that Man should be reclassified as an ape or that chimps should be re-classified as hominids than to the conclusion that our analysis of DNA variance is either not as accurate or not as meaningful as some may claim. But still the closeness of the match implies that Man and chimpanzee shared a common ancestor until a relatively short time ago.

Contrasting a human and an ape, the most obvious differences are cultural – the human is the one wielding the chainsaw and the ape is the one cowering in the tree-top. Humans wear clothes, drive cars, live in cities, play golf, shoot each other because they can't agree who was God's last messenger etc. Apes, in general, do none of these things. Physically at first glance they seem quite different too but many of these differences are superficial. For example, humans are virtually hairless where all other apes are covered in hair. However, from an archaeological point of view hair is not much help as it is very rarely preserved in the fossil record. The

70

same goes for the fact that humans have thick lips and pendulous breasts: soft tissue generally rots and leaves no trace.

There are, however, distinguishing physical characteristics between humans and apes. The most obvious differences derive from Man's habit of walking upright. An upright stance and bipedal gait are two of the things that make a human a human and an ape an ape. Both are hugely important in the development of humans and both occurred very early in their development. The upright stance frees the hand from duty as a foot. In humans the hand has become a precision tool, far more versatile than the ape hand. The opposable thumb gives a delicate but firm grip compared to the ape's more powerful but less exact hold. On the other side of the coin, walking on two feet has led to a loss of functionality in the foot department. We have lost the prehensile foot of our ancestors. While an ape can peel a banana with its toes there are few people who can do this trick. Our toes are mere nubs of more or less useless bone and gristle. The upright stance has resulted in changes to almost all the skeleton – the foot, ankle, knee, hips, back, neck and skull. All of these changes are apparent in the fossil record.

The other main areas of difference are in the head. Humans have relatively small canine teeth – we have lost our fangs. This is good for the archaeologist – teeth are made of enamel which is very hard and fossilises well. There are also differences in the palate and larynx which are associated with speech. As language is one of the main features which distinguishes humans from all other animals, these differences are very important but unfortunately we are once again dealing with soft tissue. In some cases it has been possible to construct a model of the vocal passage from fossil remains but the accuracy of any model is questionable – there is little hard evidence. The other main distinguishing features of the head are the chin (this juts out in humans), the brow ridge (humans don't have one), the general face shape (the human face is flatter) and the size of the brain pan. The chin, face-shape and brow ridge are interesting as they are features which occur in many fossil finds and show a general pattern of progression. The general face shape becomes flatter. The brow-ridge diminishes steadily from being a prominent, heavy ridge to being hardly there at all in modern humans. The chin progresses from a receding chin like a chimpanzee to the proudly jutting square jaw of the Hollywood hunk. This progression is one of the indicators as to where in the scheme of things a fossil might be placed. Of course, things are very rarely that simple, fossils do not fit neatly into a sequence. One fossil may have a more prominent brow ridge but more modern dentition or will have an archaic jaw shape but a larger brain pan.

It is with the size of the brain pan that we finally get down to the crux of the matter.

Of course the brain is soft tissue – just about as soft as tissue gets. Brains themselves are very rarely preserved in the fossil record but the size of the brain can be estimated from the size of the skull and the size of the skull can be estimated from surprisingly small bone fragments. The brain of a gibbon is around 90cc, that of a chimpanzee is somewhere around 400cc. A gorilla has a slightly larger brain at around 500cc but then again it has a much larger body. Humans have a whopping 1200 to 1450cc of grey matter tucked up inside their skull controlling a body half the size of a gorilla's.

This progression from small brain to large brain is a feature of human evolution. The first identified hominids had brains not much larger than an ape of similar size. This explosion of brain capacity happened relatively suddenly in evolutionary terms. Between about 2,000,000 years ago and 700,000 years ago the brain size doubled from about 440cc to 900cc and then increased again to its current 1,300cc by about 100,000 years ago.

By the time we get to *H. sapiens* we have a brain capacity out of all proportion to our body size. Before getting too excited about this it should be born in mind that a squirrel has a higher brain size/body size ratio than any human but it would score pretty low in an IQ test. However no animal of comparable size has a brain a big as a human. While an elephant's brain may weigh up to five kilograms it is still smaller relative to total body-weight. Not only is the human brain relatively the biggest of the animal kingdom but it is also the most complex, the most sophisticated and the busiest of them all.

A big brain needs a big head and this is a problem when giving birth, as any mother will tell you. This problem is exacerbated by changes to the hips which have become relatively narrow as one of the consequences of an upright stance. To make things easier for mum, human babies are born at a very early stage of development – this is called neotony. A human baby has a brain only a quarter of its adult size; in contrast a chimpanzee has a brain half its mature size at birth. Elephants babies by the way have a brain about one third of adult weight. Of course there is a payback for this neotony and the payback in Man's case is his extended childhood. A chimpanzee reaches physical maturity at about eight or nine years while it takes until about fourteen or fifteen years in humans.

Humans compensate for this long childhood by having a long life. Most animals' life-span is related to body size – an elephant lives longer than a dog which lives longer than a mouse – but in fact the heart of each of these will beat about the same number of times, just faster or slower. It has been calculated that most mammals live for about 1.5 billion heartbeats. For humans this would give a life-expectancy of around thirty to thirty-five years but in this as in so many other cases, we step outside the pattern of nature. Today, with improved living conditions, diet and medicine the average age of death for people living in first-world countries is over seventy years, although in the third world many die much younger.

While it is true that average life-expectancy prior to the nineteenth century was only about forty years, this is largely due to the fact that many children died very young: if you survived past twenty your chances of surviving to say, sixty, were not too much worse than today. However, it is probable that our early ancestors lived rather shorter lives than we do.

These differences in growth patterns may distinguish humans and apes in but often they may not in fact make it easier to identify humans from apes in the fossil record – a bone with a certain amount of growth may belong either to an older ape or a younger hominid.

But at the end of the day it is not bipedalism or jaw shape or brow ridges or growth rates which make a human a human: it is intelligence. Man is inordinately proud of his big brain. The fact that we can intellectually understand more of the universe than any other creature is a matter of endless self-congratulation. The estimated size of the brain of any fossil find is vital in deciding its place in the grand scheme of things. In brain-size, bigger is taken to be better.

That just about covers the physical differences.

To return to the story…

After the Ramapithecenes there is a long gap in the fossil record. What happened during this period is simply unknown because no fossils have been found (this is true at the time of writing; things might be different now). You may come across *Ardipithecus, Sivapithecus* and/or *Sahelanthropus tchadensis* and others which fit into the picture somewhere around this time, although exactly where is up for grabs. It is possibly here that *Orrorin* fits in.

The next more-or-less agreed species in the progression to Man are the Australopithecenes, although there is conjecture that the hominid line sprouts directly from *Orrorin* and that the Australopithecines are just a side issue. Be that as it may, the Australopithecine line split off from the main primate stream somewhere between 10,000,000 and 5,000,000 years ago. Australopithecenes are split into two main types – the more slightly built (gracile) and more robustly built. You may find the robust species referred to as *Paranthropus* or this may be a separate genus altogether. There were various species – *A. anmensis, A. afarensis, A. africanus, A. garhi, A. boisei* and *A. robustus* are the most usual although depending on which book you read, you may find a few more or a few less or the same number but under different names. Just how many species there were varies from year to year and from author to author.

The reason for all this confusion is that when the first hominid fossils were discovered there was a tendency for each new find to be declared a new species. This led to a proliferation of species, many of which were, on closer examination, very similar. As more fossils have been found and older ones have been re-examined, there has been a tendency to re-classify them into fewer species. These two trends – splitting and lumping – have led to a hugely confused and disputed picture. There is little agreement on the details and each new find tends to obfuscate the obscurity.

The problem is that a species is generally defined as being a population which can inter-breed and produce viable offspring. Without living examples, it is impossible to establish with any real certainty where species lines should be drawn. What makes things more difficult is that fossils are of individuals and individuals are all different. For example, if a leg bone is found that is smaller than usual, this may mean that the fossil is that of a child or small adult but if a dozen such bones are found it may imply that there was a race of small individuals which may imply a new species. Until a large number of bones are found which can be classified together, we cannot be exactly sure of what we are looking at, and the fossil record being what the fossil record is, the chances of finding many bones, let alone agreeing that they should be classified together, is slim.

A further complication is the fact that physical variation may not necessarily imply a different species. If a future archaeologist dug up the skull of a Great Dane and the skull of a Chihuahua it is unlikely these would be immediately recognised as belonging to the same species, which they are.

It is also true that finding a new species makes far bigger headlines and has far more prestige and academic plaudits, not to mention bigger research grants, than just a digging up a few more bones of something already discovered by someone else. The temptation when a new fossil is found is to claim it is as new species and let others dispute it…which of course they do.

The earlier precursors may or may not be classified as hominids depending on what criteria you choose but with the Australopithecenes we take the great leap forward into humanness. Australopithecenes are hominids; that is they are human-like rather than ape-like, although it would be pushing things to actually call them human.

The most significant Australopithecene finds have been made in east Africa. These include the famous 'Lucy' skeleton, the 'Taung Child' and 'Nutcracker Man'. There is also the Laetoli pavement.

The Laetoli pavement is a fossilised set of footprints left by a group of *A. afarensis* as they walked across a patch of volcanic ash some 3.8 million years ago. If the ageing is correct, this implies that Australopithecenes had adopted the human, upright bipedal gait all that time ago. I should emphasise that simply being able to walk upright does not make an animal a human – a dog can walk upright for short periods as can a chimpanzee. But examination of the tracks – and they have been examined, analysed, re-examined and re-analysed ad nauseum – shows conclusively that these animals (people?) habitually walked upright. They had Man's upright stance and bipedal gait which, ipso facto, makes them hominid. The creatures walking in the Laetoli ash would look more ape-like than human-like to most modern humans – they were more of an ape's size, probably hairy and had none of the usual human accoutrements such as clothes or weapons. They had brains not much bigger than a chimpanzee. But they walked upright as no ape had done before them.

Modern apes have various modes of locomotion. Gibbons swing through the trees using their long arms – this is known as brachiating. Chimpanzees may do the same but on the ground they tend to walk on all fours using the knuckles of their front legs. Gorillas are too big for brachiating and generally knuckle-walk like chimps. Orangutans use a mixture of methods.

Why the Laetoli Australopithicenes walked upright is a good question and one which has been the subject of endless speculation. By this time climate change had altered the landscape such that the dense forests where these creatures lived had been largely supplanted by semi-open country. The upright stance would allow them to see further and to pick berries from the trees with their hands – but an ape can stand upright to do this without walking. It has been argued that it is a more energy efficient means of locomotion or that it aids heat dissipation but this is far from certain – it has equally convincingly been argued that it is less efficient and more prone to accident. The upright stance developed several millions of years before the first known use of tools so it was probably not done in order to free the hand. It has been suggested that the origins upright stance may be seen in the orangutan's habit of walking along branches while holding on with their long arms.

Neo-Darwinists have an obsession with finding 'the reason' why some evolutionary development occurred. As humans are taken to be the most successful species ever to exist then it is assumed that any feature which make humans distinctly human must give some evolutionary advantage. This is questionable on several levels. Firstly as mentioned above, humans should not necessarily be regarded as an evolutionary success. Secondly there is really no reason to regard humans as any more successful than the apes that continue to knuckle-walk or brachiate. Thirdly the upright bipedal stance may not be the advantage – it may simply be a concomitant development of some other mutation which was advantageous – more frequent mating for example. Furthermore it is possible that bipedalism could be a 'pre-adaptation' which in itself conferred no advantage but later, with further development of the hand and brain, became advantageous

There may be no reason. In neo-Darwinist terms, a random mutation occurred which resulted in changes to the foot, leg, hip, spine, neck, skull and behaviour of our distant ancestors with the result that they walked upright. This did not cause sufficient disadvantage to these creatures for them to die out and so this mutation has up to this point been preserved by natural selection.

The Australopithecenes flourished between about 4,000,000 years ago and 1,000,000 years ago. Some time around 2,400,000 years ago a new genus arose: *Homo*. First came *H. habilis*. Some time later came *H. ergaster* and about 600,000 years ago we get *H. heidelbergensis*. The earliest claimed fossil of our own line, *Homo sapiens* was found recently in Ethiopia and dated at around 200,000 years ago although this is exceptionally old. Even more recent even older finds in Israel just confuse the issue even more. You may also find *H. rudolfensis, H. erectus, H. antecessor* and a few others somewhere in the mix. Neanderthals (*H. neanderthalensis*) were around between perhaps 120,000 years ago and died out in about 30,000 years ago. Where exactly the recently discovered so-called 'Hobbit' (*H. floresiensis*) fits into the picture is an open question – the most modern fossils found date from only about 15,000 years ago. About 50,000 years ago the star of the show, Cro-Magnon Man, turned up. This is modern Man, modern *Homo sapiens,* you and me.

The succession of species is not a neat and clear-cut matter of one form superseding another, there is considerable overlap. Even the number of distinct species in not agreed. Some species co-existed at the same time and whether these species interacted and if so how and how much is a matter of conjecture. Whether the interaction went as far as inter-breeding is another vexed question. DNA evidence suggests that the Neanderthals split off from the main line and are not one of the direct ancestors of Man but this is far from proven. There is some speculation that rather than being replaced by *H. sapiens* they were absorbed by inter-breeding although they seem to have contributed little if anything to the modern human line. Perhaps the ongoing human genome project will shed more light on this. There certainly seems to have been contact between the species as there are striking cultural and technological similarities. The ranges of both appear similar although the Neanderthals appear to have occupied more northerly regions before being supplanted by *H. sapiens*. Both share similar tool technologies although the Neanderthals did not reach quite the degree of sophistication of *H. sapiens*. Fire, clothing and burial practices appear to be common to both.

I will not even try to list the physical differences between species but the stance, jaw, teeth and skull are still the major physical delimiters. Basically the progression is as before – refinement of the facial features, disappearance of the prominent brow ridge, development of a squarer, more jutting chin, changes to dentition and enlargement of the brain.

However, in most cases it is not simply the bones that tell the story.

Archaeologically speaking, when a possible hominid site is discovered there are more indications of where it may fit in the

79

family tree than simply fossil bones. Often of more significance is the trash that has been left behind. It is a telling fact that, from the very start, hominid sites should be identifiable from carelessly discarded tools, from the waste from meals and from environmental impact. It is this clear distinction between natural and man-made which allows us to identify one site as hominid and another as 'natural'. Many, probably most, sites have been identified in the first instance from middens (rubbish heaps) or from tools found strewn about.

The use of tools is the hallmark of humans. Tool-using used to be considered the exclusive province of humans but it is not: sea otters have been observed to use rocks to crack shells and so have some species of bird; chimps may use a twig to search for termites. Chimpanzees have even been known to prepare their sticks by stripping leaves and bark to make a more effective tool. However tool use in humans reaches a level of sophistication which is unique. What is more, humans are the only animals to use meta-tools – tools for making tools. One of the most significant changes in early stone tool-making came about when people started using stone hammers and chisels to sharpen the edge of other tools. Tools of one sort or another are found at every human site. They are an identifying feature, indeed a required feature, for the site to be designated human.

The earliest identified tools are over 2,000,000 years old. They are associated with *H. habilis* sites in East Africa and consist of small, sharp flakes of quartz plus some larger, roughly-shaped pieces and pebbles. There is some indication that even older tools, perhaps up to 2,500,000 years old, may have been used by some of the robust Australopithecenes but this, as so much else, is disputed.

From the bones and other debris found with the *H. habilis* tools it looks like they were used for preparing food. *Homo* by this stage had lost the fangs and strong jaws of their pongid ancestors and yet evidence from analysis of bones and tooth wear suggest that they had moved from a straight vegetarian diet to a more omnivorous diet including some meat. With their reduced teeth they needed some help in ripping through the tough hide of the animals they preyed on, tearing the meat from the bones and crushing those bones for their nourishing marrow, so they invented tools. No other animal could or would do this. Most carnivores have big, sharp teeth and strong, crushing jaws which are well adapted (naturally selected) for flesh eating. Most animals with teeth adapted (naturally selected) for eating vegetables, eat vegetables. *Homo* neither had the teeth for meat nor was he content to remain a vegetarian: he went in a new direction.

There has been plenty of speculation about exactly how these early people may have lived. The answer may have more to do with the social and political climate prevalent when the question is asked than with the physical evidence. In the nineteenth and much of the twentieth century the answer was that Man was a brave and noble hunter who ventured out to provide the meat for his family while the womenfolk gathered what they could from the countryside. The tools were for Man to kill and butcher his prey. Now, with a less sexist and a less rosy view of Man, early Man is seen more as a scavenger of carnivore kills who brought back his offerings to his womenfolk for processing. The physical evidence is tenuous and far from certain. The scavenging interpretation comes from analysis of breakages and scratch marks on fossilised bones found with the tools and the fact that, at this stage, there is no unequivocal evidence of spears or other weapons for actually killing prey. On the other hand, wooden spears or weapons would not be preserved in the fossil record and they may have been in use even at this early stage. It is probable that the truth lies somewhere in the middle. Man is the ultimate opportunist and if the opportunity arose for scavenging then no doubt he would have scavenged but there is little reason to doubt that if the opportunity arose for killing, he would have killed. If there was fruit, he would have eaten fruit, or seeds, or insects or whatever else presented itself.

There is some speculation that Australopithecenes or early *Homo* could have been cannibalistic. Again the evidence is uncertain but it does suggest that there have been instances of cannibalism in the human story although to what extent and whether it was ritualistic or not are still debated.

Man's tools show a general progress from the crude, roughly shaped tools of *H. habilis* through the better formed and more specialised tools of *H. erectus* to the sophistication and sheer beauty of the tools of the late Stone Age which are well designed, finely crafted and sometimes even decorated. One interesting point on this is that we see progress in leaps and bounds – the tools do not gradually improve from form to form but seem to be either one form or another with little middle ground. We should perhaps not be surprised by this – this appears to be a characteristic of Man. Man seems to doddle along for a bit and then he reaches a point where there is sudden change. There have been several of these sudden jumps – the Upper Palaeolithic revolution (we're coming to this), the change from hunter-gatherer to farmer, the Renaissance. In the last few hundred years we have had the agricultural revolution, the industrial revolution and we may be undergoing another revolution right now. These revolutions have been sudden and pervasive and seem to be general. They do not start somewhere and spread but seem to arise all over, all at once. What is more the pace of change is increasing. The oldest stone tool industry – the so-called Oldowan industry – persisted largely unchanged for a million years. Then there were the improved Acheulean tools which were superseded by the Mousterian. From the late Stone Age onward the pace of change has been fast and is getting faster. It took perhaps 50,000 years for Man to learn to fly but less than one hundred years to progress from the Wright brothers' flyer to the moon program

Innovations appear to appear all over, all at once. This pervasiveness applies at both the macro level with large-scale movements such as settlement and the development of agriculture and at the micro level with such specific inventions as fish hooks and pottery. These movements appear to have happened to many people in many places at approximately the same time. Take for example fire. There are early hints of the use of fire going back many hundreds of thousands of years but its use becomes pervasive in the last 100,000 years. When I say pervasive I mean that suddenly everyone is using fire, from South Africa to Europe to China. It is of course possible that fire was tamed in one place at one time by one person and that this knowledge was then disseminated throughout the globe. There has even been conjecture that perhaps an alien spaceship landed and taught everybody how to use fire but both of these scenarios raise more questions than they answer and seem unlikely. This same characteristic of sudden, pervasive appearance goes for settlement, clothing, burial, agriculture...the list goes on.

The use of fire is one of the distinguishing features of humans. The earliest unequivocal evidence of the use of controlled fire comes from about 500,000 years ago, the time of *H. ergaster/erectus*, although there are tantalising hints of earlier use, perhaps even as early as 1.5 million years ago. Exactly how Man first controlled fire and when he first learnt to make his own fire is unknown but the impact of human-controlled fire is indisputable: as Stephen Pyne puts it in 'World Fire': 'Fire and humanity have become inseparable and indispensable. Together they have repeatedly remade the Earth'.

The advantages of controlled fire are obvious – heat on a cold night, light in the dark, protection from other animals, preparation of food. Humans are unique in their use of fire. Most animals shun fire: humans see it as another tool.

When people started wearing clothes is impossible to say but it was probably early on. Hard evidence is rare because clothes are mostly made of animal skins or plant fibres and neither of these materials preserves well. What is more, the first clothes were probably no more than animal skins thrown around the shoulders so they may not be recognised unambiguously as clothes anyway. By about 40,000 years ago there are bone needles and what look like leather-working tools so there was sophisticated, sewn clothing by this time but the first person to wrap a fur around themselves must have lived many thousands of years before.

Clothes do more than keep you warm – they insulate you from the environment; they protect your modesty; they make a fashion statement. No animal apart from Man wears clothes; they are content to go around 'as nature intended'. Some animals living in colder climates have developed mechanisms to deal with the cold such as developing a layer of blubber beneath the skin or a thicker winter coat or making their fur stand erect. Not so humans. Without fire and clothes Man's range would be severely restricted. We would be forced to live in the temperate zones where our puny bodies could cope with the weather. But we have our clothes and so we can live wherever we please.

With tools, fire and clothes we begin to see the separation of humans from the natural world. They allow us to keep warm without being covered in hair, to see in the dark, to soften our meat so that even our poor teeth can manage it. Man made the giant leap from being dominated by his environment to being the one who dominates at the earliest stage of his existence. We circumvented the natural constraints of climate and diet which might keep another species in check. Our use of tools, clothes and fire were not evolutionary modifications which made us more adapted to our environment – in Darwinian terms, fitter – they were, on the contrary, innovations which allowed us to leave our normal environment and explore other environments. We have no need of a cat's night-vision, its fur or its sharp teeth – we have our fire, our clothes and our tools. Man has climbed out of his niche.

He has also moved beyond natural selection. Man does not have any niche and therefore cannot be fit for it. If it gets colder (which it did through several Ice Ages) then he simply puts another log on the fire and pulls another skin around him. If a child was born with a thick subcutaneous fat layer or longer, thicker hair giving a better natural tolerance to cold, who cares? This gives no advantage at all over his thin-skinned, naked brother or sister sitting by the fire in their bear skin. It might even be a disadvantage depending on whether being fat and hairy is taken to be sexually attractive or not. The Neanderthals appear to have been better adapted to the cold – they had wider noses to warm the air they breathed and a more compact body shape – but this did them no long-term good in their struggle for life. Cultural selection overtook natural selection as the primary evolutionary driving force very early in Man's development.

At some point between about 500,000 and 50,000 years ago Man came up with what is probably his greatest ever innovation: language. Exactly where, when, how and why language developed is a matter of conjecture and endless debate. Unfortunately the physical evidence is sparse and ambiguous as the parts of the body most involved in speech – the larynx, tongue, soft palate and brain – are all soft tissue which is rarely preserved. There are two specific areas of the brain – Broca's area and Wernicke's area – which are associated with language and much research has been done trying to work out how developed these particular areas are in the sequence of hominid fossils in an attempt to pin-point the origin of language. So far results are inconclusive. This leaves only circumstantial evidence and inference as vague pointers to the origins of language.

It now seems that a particular gene – so-called Foxp2 – is critical in vocalisation in humans and in animals. This gene first came to light in studies of a family with varying degrees of speech defects. It was found that all members of the family who had inherited a mutation in this gene from their maternal grandmother, had speech defects and those with no mutation had normal speech. Since this time extensive studies have been done and it has been suggested that mutation of this gene in humans coincides with the emergence of modern *sapiens*, although this conclusion is far from certain.

It may well be that the propensity for language is in-built in the human brain. There have been efforts made to analyse the 'deep structure' of language and prove that this is common to all peoples at all times although the results are inconclusive. Certainly some languages are related, for example the Indo-European group which comprises many of the languages of India and Europe. However attempts to establish relationships between the major language groups and infer some sort of common, original proto-language have, so far, been unsuccessful. It is quite probable that language arose independently more than once and in more than one place.

There are many theories – too many to consider in detail here – and new theories, new evidence and new conjectures are being constantly added to the mix. The truth may never be known.

We take language so much for granted that it is impossible to envisage a human society without language. Chimpanzees use some vocalisations but these do not come close to what we think of as language. Experiments with chimpanzees and bonobos have proved that they can learn to sign or use symbols to some degree but the most sophisticated chimp, even after years of training, has never progressed much beyond the level of language achieved without effort by any five-year-old child. These experiments, while interesting, really prove absolutely nothing about a chimp's 'natural' propensity for language. Language reflects Man's world-view and we have no reason to believe that a chimpanzee views the world in the same way. Making chimps use human speech to try to get an insight into how language developed is like teaching them to wear clothes in order to get an insight into our fashion sense.

Language is another attribute which used to be regarded as exclusively human but this is questionable, depending on how you define 'language'. Birds have their calls which are certainly a form of communication; some animals have alarm calls and some of these are specific, for example vervet monkeys use one call to warn of an approaching snake while they use a different call to indicate a hawk. Then there are also such enigmas as whale-song, the clicks and squeaks of dolphins, the howling of wolves and the flashing colours of cuttlefish which may or may not be communication and which may or may not be understandable as language. However, Man has taken language to a level of sophistication unparalleled in the natural world.

Human languages, all languages, have three characteristics which make them different form all other forms of communication: productivity, duality of structure and displacement. Productivity means that although there are only a finite number of letters and words these can be combined to make up a virtually infinite number of sentences. Duality of structure means that elements of the language may be put together in different ways to express different things. For example 'John hits Jim' and 'Jim hits John' use the same words but mean two different things. Most importantly, displacement allows us to talk of places, times and conditions other than where we actually are.

There are many, many different languages in the world today. All cultures have a language and children will naturally learn the language of whichever culture they happen to be brought up in – an American baby brought up by Kalahari bushmen will learn bushman language without any special effort or instruction. What is more, all languages are translatable into all other languages. There are differences in nuance and some translation may need to use a good deal of circumlocution to achieve a result but basically anything you can say in any language, you can say in any other language.

The most obvious use of language is as a medium for passing and receiving information and this aspect is generally taken to be its primary function and probably the motivation for its original development. One theory is that language developed from cooperative hunting. However wolves, lions, even pelicans hunt cooperatively and do so without (as far as we know) using language. Also hunting is generally a game of stealth and quiet – even among human hunters communication is more often by hand signals or other non-verbal signs than by spoken words.

But actually, in everyday conversation, much of what we say is not informational as much as conversational. In many cases a human conversation serves much the same purpose as mutual grooming in less vocal species. Perhaps language developed simply to fill up that awkward silence that sometimes fell around the campfire.

Language allows the communication of abstract ideas and concepts. It is arguable that abstract ideas and concepts are only possible within the context of language and there has been endless philosophising on the relationship of language and thought without any final conclusions. We are so imbued with language that it is hard to think without verbalising, if only internally. Language allows Man to pick up the wisdom of generations simply by listening to his elders and to pass his experience on for the future. It allows Man to tell stories, to joke and to lie – it is doubtful that a chimpanzee can do any of these.

But again with language we see a separation of Man from nature. Most animals appear to live in the here and now, the eternal present: language allows Man to live wherever and whenever. We spend much of our time reminiscing about the past or planning for the future. We can muse and speculate, wonder and conjecture about things which do not exist or of which we have no experience. It is almost impossible to over-state the importance of language in Man's development and in the speed of that development. It is central to the development of what is to come.

Of course it may be that a may-fly or a salmon or a gnu also speculates about the meaning of life but, given their lifestyles, it seems less than likely.

With the emergence of modern *H. sapiens* about 50,000 years ago, there is a leap in development. Tools suddenly explode in number and quality. Instead of the simple, crude, utilitarian artefacts of early Man we suddenly find finely worked, precision instruments, some with decoration. New materials such as bone and antler are utilised. Musical instruments first appear. There is a great increase in mural and portable art and evidence of a generally greater degree of sophistication in lifestyle.

The first conjectured ritual burials appear about 100,000 years ago and are associated with Neanderthals and burial is certainly a feature of many later hominid sites. Usually the body or bodies are carefully placed and in some instances traces of pollen have been found suggesting that flowers may have been placed in the grave. In some later burials the corpse is decorated with jewellery or other artefacts. What are we to make of such behaviour? There is no precedent in the animal kingdom. Elephants have been observed to hang around and touch a dead companion with their trunks or to play with an elephant skull so they might have some concept of death but ritual burial implies an awareness and awe of death, perhaps even belief in some sort of life after death, which is beyond the ken of any animal.

From about 35,000 years ago to maybe 15,000 years ago was the era of cave painting. When the first cave art was discovered in Altamira in Spain, people could not believe that it could be as old as it was claimed. It was so 'modern' in its appearance, so sophisticated. The question most often asked about cave painting is why? The most common answer used to be for magic or shamanism – the animals were drawn to give the drawer some sort of power over them but this does not seem to fit the facts. The number of drawings of animals does not correspond to the number of bones found in the caves. The most common prey animals are not the most frequently depicted. The paintings are taken to be more than simple wallpaper to brighten up a dull old cave. Many of the paintings are deep underground in inaccessible corners and it would have taken considerable effort to produce them. Are they just the idle graffiti of some bored caveman waiting for his supper? There are some scratches in more accessible caves which look very much as if they are just that but even the hidden works could be no more than art for art's sake, the urge to leave one's mark, the 'Kilroy was here' of the Upper Palaeolithic. At the end of the day, why does any contemporary artist paint and why should the motivation be any different just because the artist lived 30,000 years ago? The question of what these paintings 'mean' I will leave to the pundits but, when all is said and done, they are decoration: Man has started decorating, permanently altering, 'improving' his world.

Several carved statuettes have also been found dating from this time. These are mainly small figurines of exaggerated femininity and they are mostly made from ivory. These may also have some religious significance or be fertility symbols but, when all is said and done, they are ivory statuettes. This means that a mammoth died so that they could be carved. Perhaps the mammoth died of natural causes and the ivory was scavenged. Perhaps...

There is sometimes talk of a turning point in Man's development. Was it the use of tools, or fire or clothes or the development of language or awareness of death that first set Man apart? There is no single turning point, just a gradual but inexorable trend away from the mainstream. No single innovation makes humans human, it is a whole package.

The spread of Man around the globe is another source of contention. The current most popular thesis is the 'Out of Africa' theory. Archaeological evidence and genetic research both point to an origin in east Africa. The genetic evidence is based on mitochondrial DNA. Mitochondrial DNA, as distinct from nuclear DNA, is DNA found in the body of the cell rather than the nucleus and it is passed on from the mother only, not from both parents. Small, insignificant changes occur in this DNA which accumulate over long periods of time into what is known as 'genetic drift'. The rate of these changes appears to be fairly constant so by analysing the number of differences between two samples it is possible to estimate how far back in time they shared a common ancestor (mother). Comparing DNA of modern humans using this method it has been suggested that all living humans are related back to a small number of individuals, perhaps just to a single female, so-called mitochondrial Eve, who lived in Africa some 200,000 years ago. If this conjecture is correct then all living humans are the eventual progeny this very small, related group. One implication of this is that all the investigation and speculation surrounding Man's deep origins may be pretty much irrelevant. If we all derive from a few individuals then it may be that these were an exceptional group.

The alternative thesis: that modern Man evolved gradually from older *Homo* species all around the globe is no longer much supported. The question then arises: are we the exclusive descendants of this original group (the 'replacement' thesis) or did others inter-breed with this group and if so who and how much. Both theories have their advocates and the evidence is ambiguous enough not to exclude either.

From Africa we spread across the globe. There are currently (2011) approximately 7,000,000,000 people in the world. The DNA samples used in the original Eve hypothesis came from only 147 people, a tiny percentage of this population so the conclusion is by no means certain and, as with most theories concerning Man, it is disputed. However larger samples have yielded similar results and the fossil evidence tends to support, or at least not to contradict the thesis. There is currently a project under way to analyse large numbers of DNA samples from around the globe in an attempt to map Man's progress. It will be interesting to see how this agrees or disagrees with current theories – so far all signs point to it supporting the thesis.

Why Man spread is a matter of conjecture – perhaps it was climatic influence or population pressures or pursuing game – but the explanation could be as simple as curiosity. Man has always wondered what is over the horizon. Man has never been content to let things stay as they are. Man needs change.

Man spread rapidly. By 150,000 years ago Man had expanded into the steppes of central Russia. There is evidence of humans in Australia by about 50,000 years ago although there is some question as to how exactly they got there across at least sixty kilometres of open water – there were no comparable sea voyages before nor for many millennia after. America was invaded from Asia via a land bridge across the Bering Sea probably about 35,000 years ago but the main invasion came only about 11,000 years ago. These so-called Clovis people spread like a plague in their new found land, supplanting the older Fulsom culture and reaching the Gulf of Mexico in only a few hundred years. By 10,500 years ago they had reached the southernmost regions of South America and numbered in the millions. As the long Ice Age came to a gradual end about 12,000 years ago, Man followed the retreating ice and by 10,000 years ago there were people living within the Arctic Circle.

As Man expanded, so the megafauna (mammoths etc) disappeared. In North America for instance, thirty nine genera of megafauna went extinct by about 11,000 years ago, including horses and giant ground sloths as well as mammoths and rhinos. A genus (plural: genera) is a group like apes or rhinoceroses. In Eurasia a similar number of genera went extinct at about the same time. In Australia, 85% of all large animals disappeared. Africa got away lightest with only eight extinct genera. Is this a coincidence? This is a thorny question. The evidence is ambiguous and can be interpreted in different ways. There were major climate changes which had profound effects of the environment and would have reduced the range of the megafauna anyway.

However, in speculating about the behaviour of an ancient animal, it is often instructive to look at a similar animal alive today and see if their behaviour may provide any clues. Thus we might expect a mammoth or a mastodon to behave somewhat the same as a modern elephant, taking into account changes of climate, vegetation etc. We can say with certainty that Man right now is wiping out species on a daily basis. The evidence of extinction following occupation by Man is undeniable. New Zealand was invaded about 1,000 years ago by Polynesians. Since that time 50% of bird species including all the large flightless birds have become extinct and climate change is not to blame. A similar story can be told for Hawaii and most other islands.

While there have been dramatic climate changes during Man's early history – there were three Ice Ages in North America between 70,000 and 10,000 years ago – there are still ice-caps and there is still plenty of tundra not at all dissimilar to conditions when woolly mammoths and rhinoceroses thrived. Perhaps numbers dwindled with the ending of the Ice Ages but Man was there to finish off the job. There is little doubt that Man was, at the very least, a contributing factor in the extinction of the megafauna.

The Neanderthals also disappeared at around the same time. Whether *H. sapiens* did a similar hatchet job on the Neanderthals is open to debate. Often the answer is influenced as much by the debater's view of *H. sapiens* as by any objective facts. With the amount of conflict today within such a homogenous human race, we may reasonably conjecture that *H. sapiens* and Neanderthals may not have got on well together, but there is no evidence of killing fields or mass slaughter which are the usual hallmarks of genocide. If the Neanderthals were exterminated by *H. sapiens* it may have been that they were simply superseded by a 'fitter' species by gradual degrees and with no particular malice. Or it could have been that they suffered the same fate as many native cultures suffered in the Age of Empires when the Europeans came: pushed aside, infected with alien diseases, enslaved or killed. The Neanderthals were an obstacle to Man's expansion and that is a dangerous thing to be.

Man versus Life

Man is wiping out species at an alarming rate. From the megafauna onwards we have been killing directly and indirectly by destroying habitation, usurping resources and polluting the environment.

There is a wilfulness about this that is frightening. The continuing killing of rhinoceros, tigers, and so many other animals when even the poachers know that numbers are dwindling is incredibly short sighted. In 1874 the passenger pigeon was one of the most common birds in America. They flew 'in such compact masses that they absolutely intercept the rays of the sun'. By 1900 they were hunted to extinction. The dodo was wiped out by bored and hungry sailors. The Tasmanian tiger was wiped out by paranoid farmers. In America the beaver and the bison both numbered in tens of millions but were still nearly wiped out for fashion, food and sport.

Once a species is extinct it is gone forever. Even if we could clone, say, a Tasmanian tiger from genetic material we have, what would the resulting animal be? It would come into a world knowing nothing and with nobody to teach it to be a Tasmanian tiger. It would be a circus freak – a clever trick but nothing more. You cannot resurrect the dead.

There are several species on the edge of extinction now where most of the survivors are in zoos in the fond hope that the global trend of expansion will one day stop and they can be released again to the wild. This seems to be based on optimism rather than any real likelihood that appropriate wild areas will somehow become available. It seems inevitable that, barring some unforeseen event like a comet, Man's population will continue to expand, more habitats will be destroyed, and more species will be wiped out.

But even if Man has opted out of nature, nature has not necessarily opted out of Man. Viruses are evolving too. AIDS is a world-wide issue now and will continue to grow in impact unless a cure is found. About a hundred years ago, a 'flu epidemic killed 20,000,000 people. The recent SARS, Bird Flu and Swine Flu scares show the potential of viruses to mutate quickly into killer strains against which the human immune system is ineffective and it is all but inevitable that new mutations will appear. There is no doubt that new viral strains are as great a potential danger as climate change or meteor impact. Man may have wiped out smallpox but some remote descendant could still return the favour.

Life is resilient even if species are not. Species come and go and if Man is wiping out a few now, even if he is the agent of the sixth extinction, then life will go on. Man has had an impact but it too will pass and the Earth will recover.

Settle Down and Raise a Family

Easter Island is one of the most remote inhabited places on Earth. It is in the eastern Pacific Ocean over 2,000 miles from the coast of South America and a similar distance from Tahiti, the nearest landfall to the west. Easter Island is known to most people who know it for the many massive carved stone heads which are scattered about the island. These enigmatic sculptures are pretty much the only things of note on the island: apart from them it is a bleak and barren place – few trees, little wild-life, few people.

Easter Island was not always like this. A few short hundred years ago it was covered in rainforest with all the usual accompanying bird and animal life. And then in about 400 AD it was discovered by Man and his accompanying dogs and goats and pigs and rats. These were not rapacious Europeans bent on pillage but Polynesians – Rousseau's 'noble savages' – come to populate another stone in the ocean. They cut down the forests, hunted the animals, carved the sculptures and developed a complex culture. The population grew eventually to outstrip the scant resources of the island. The goats and pigs prevented the re-growth of the forests, the dogs and rats killed the birds and small game. For some centuries the civilisation flourished but population pressure on the diminished resources led to friction and to virtually continual, almost ritualised, warfare. Eventually the population was reduced to remnants and the civilisation foundered.

Nowadays, tourists fly to the island to admire and wonder at the massive statues, but Man's legacy is not just his art, it is the whole island. Easter Island has never recovered from Man. The rainforests have not regenerated, the birds and animals have not returned. Man has permanently changed the appearance and ecology of the island and not for the better.

The story of Easter Island is to a great extent the story of civilisation. It is a story of continual, relentless expansion, of environmental rape and pillage, of destruction and extinction. It is a story of the comings and goings of empires, of war, of genocide and slaughter, of art and culture.

It is common these days to decry the state of the world, to bemoan the environmental vandalism we see around us – the destruction of rainforests and pollution – and to long for a simpler time when such problems did not exist. Get back to basics, people say, stop the capitalists, the corporates, the military/industrial complex, and return to a more simple lifestyle like, say, the Polynesians. This is an illusion. From the very start Man has been destructive. Being out of step with nature Man has tended to stomp where he will. Environmental destruction began with the first Man to cut down a tree for fuel or to make a spear or just because it was in his way, and it goes on today in exactly the same casual manner.

But we are getting ahead of ourselves. We should start somewhere nearer the beginning.

People have to eat and to eat they need food. For early Man, food was probably anything he could eat and with his all-purpose dentures and his fire he could eat just about anything. Thus it seems probable that the earliest hunter-gatherers would have been semi-nomadic, following animal herds or fruit in seasons. The number and variety of bones and tools found at some sites implies extended occupation but these may have been seasonal sites occupied over many years but for only a few weeks each year when game was plentiful. In general early Man would have lived much the same sort of lifestyle as a troop of baboons does today.

Another factor is water. Early Man would have had to remain within about a day's walk of some water source until he could carry water with him. The first man or woman to discover that you could make a bucket from an animal skin or that a hollow gourd could hold water broke a shackle that had restricted Man's range and let him loose upon the world.

However, there are signs from an early stage that some sites were at least semi-permanent and increasingly with the Upper Palaeolithic revolution and the appearance of cave art there are indisputable signs of permanent settlement.

Most animals do not settle down in the same way as Man. Some animals may have a home range which they have inhabited since time immemorial and some birds may return to the same nest site year after year but nobody (except perhaps termites) builds permanent structures like Man. Chimpanzees may build themselves a nest at night but this, like a bird's nest, will unravel and disappear without trace over time into the general maze of nature. Man's sites are still recognisable thousands of years after they were used because Man's sign does not fade as fast. He leaves his detritus – his discarded tools, the bones of his victims – and in many cases he has left a permanent mark on the landscape in the form of his art or some change to the natural habitat.

Settlement is hugely important. As a nomad you must carry everything you own with you so it behooves you to travel light. This means few or no possessions and a small family with no more children unable to walk than you can comfortably carry. Settlement changes all this. With a permanent home you can have many more possessions and they can be bigger. You can have a large extended family as grand-parents and babies can stay at home in relative comfort while the hunters go out hunting and the gatherers go out gathering.

This brings up two very significant developments – ownership and division of labour. Ownership is so intrinsic that we take it for granted – this is *my* hand, typing on *my* keyboard – but it is not so obvious. It involves viewing myself primarily as an individual rather than as part of a larger unit like a herd or a pack or a tribe or the environment. We do this from a very early age. A young child seeing itself in a mirror will quickly recognise the reflection as its own image. By the age of about eighteen months most children have this self identification – they recognise their mirror image, they start identifying themselves in speech using 'I' or 'me' or their name and they understand the concept of ownership. Whether other animals have this sense of self and self importance is harder to establish.

In 1970 Gordon Gallup carried out experiments with chimpanzees in which he anaesthetised the chimps and then put a mark on their forehead above the eyebrow where it could not be seen, felt or smelt. When he reintroduced a mirror, the chimpanzees touched the marks on their own heads, indicating that they recognised 'themselves' in the mirror. The same experiment has been tried with monkeys of all sorts, all of which fail to recognise themselves, with orangutans who usually do recognise themselves and with gorillas who consistently don't. Outside of primates, elephants and dolphins have both been tested but the results are ambiguous and inconclusive. Most other species which have been tested fail to recognise themselves. This is not necessarily because the animals do not understand what a mirror does – some monkeys have been seen to watch their keeper approach in the mirror with food – but they treat their reflection either as another animal or they do not recognise it at all. The fact that most species other than chimpanzees do not recognise an image in a mirror as being a reflection of their own body implies that they probably do not have a sense of 'self' anything like that of Man. However, the fact that many chimpanzees can recognise themselves does not in any way imply that they do have the same sense of 'self' as Man.

Taking this idea of individuality one step further, Man has developed what is known as Theory of Mind – the ability to envisage the world from someone else's point of view. Children generally develop this ability by the age of about three – they realise that even though they know something, there may be others who do not. As we get older we get more sophisticated in this to the point that we can enjoy a play where the plot revolves around person A knowing something person B does not and we can understand both points of view. Experimental results with chimps and other animals are ambiguous. Theory of mind has not been certainly demonstrated in any animal other than humans although to be fair it is difficult to see how it could be proved without language.

As most animals do not have possessions the question of ownership does not arise. In some experiments, it has proven possible to induce a chimp to hide a banana from others in the troop which may imply a sense of ownership but this is a man-made, artificial situation. In the wild, if one chimp has a banana it is probably because the group has found a banana palm so the whole group has bananas. Dogs may fight over bones or lions over a kill but this is for immediate possession, not ownership. If a pride of lions are chased off a kill by hyenas then that is the end of the story – they do not take the matter to court. The kill never 'belonged' to the lions and it does not belong to the hyenas: it is simply food to be eaten by whoever can grab a bite.

In the wild many predators will cooperate in the hunt and share the spoils, even taking food back to animals left in the home den. This sort of altruism is a vexed question for strict neo-Darwinists. If the world is all about struggle and competition and survival then the naïve assumption is that it would be the most selfish, ego-centric and ruthless who would be the 'winners' – good guys finish last.

This is no more than a hypocritical justification for human greed, disguised as pseudo-science. It is simply not true in the real world. However, instead of changing the paradigm to fit the facts, it is more usual to search for a selfish reason for unselfishness, to try to prove that altruism is in fact egotism in disguise, for example suggesting that feeding the young is simply a strategy for ensuring the survival of the parents' DNA. These types of arguments are unconvincing.

Personal ownership brings with it many more of these abstract concepts that newly acquired language is so good at. Once you have ownership, you have difference. One person owns a better spear or more furs or finer tools – he is rich – another owns nothing. We have inadvertently invented poverty. And when a rich person dies, what happens to his possessions? We can invent inheritance. And theft. And soon there will be barter and then money and banks and super-tax.

As for division of labour: there have been thousands of words wasted on this subject. Who did what and which functions are most important are questions that are unanswerable. The answers given usually depend on the prevailing political climate at the time and on the answerer's prejudices and are often inferred from the flimsiest of evidence. In the nineteenth century when Man ruled the roost and woman was seen as a meek adjunct, it was assumed that the big brave cavemen went out hunting for the main course, good red meat, while the mousy little womenfolk pattered about picking berries and such for dessert. With the rise of the women's movement this view has changed and more equal roles are now assigned to the sexes.

The first settlements were probably in caves. With fire, all you need is some sort of shelter and you are home. The earliest settlers may have been fishermen. A permanent and seemingly inexhaustible source of good nourishment right at hand along with fresh water made lakes and riversides an obvious place for early settlement.

The first definite evidence we have of construction is from the steppes of Russia where trees were scarce but mammoths were plentiful. There are mammoth bone houses dating from somewhere around 20,000 years ago consisting of the bones and tusks of many animals. These were probably cemented with mud and covered in hides. Of course mammoth bones preserve very well so the fact that these are preserved does not imply that our ancestors lived exclusively within the skeletons of dead animals. There were probably other built structures but these were probably made of wood and wood does not preserve well.

Do the mammoth bone houses imply wholesale slaughter of mammoths? Not necessarily. The bones are of various ages and may have been scavenged and gathered from miles around. The problem with this thesis is what killed the mammoths if not Man? Mammoths were the biggest things around and they had few predators other than Man. Perhaps Man waited patiently for one to die of old age before scavenging some food for his family, a few bones for his house and a tusk or two to carve on the long winters' nights.

On the other hand, when one considers, for example, the depletion of whale stocks, the on-going poaching of elephants for a scrap of ivory, the near eradication of the bison in America in the nineteenth century, the concept of indiscriminate slaughter by humans does not seem totally implausible. There are documented instances where whole herds of horses and mammoths have been found dead at the foot of cliffs in both Europe and America and there is little question as to whether they fell or they were pushed. Man seems to have been as wasteful in the cave as he is in the penthouse, driving hundreds of animals to their deaths just so that he could eat one or two before they rotted. This technique is vaunted in some books as an example of Man's intelligence and ingenuity in hunting large, potentially dangerous animals.

Once Man began to settle the door was open to new innovations. The most significant is agriculture. Agriculture appears to have appeared in several different, unrelated places at different times but all within a few thousand years of each other. There is no evidence to suggest that farming or domestication techniques were communicated directly between early communities – that Mesopotamia in modern Iraq was in contact with Tehuacan in Central America, for example – yet in both places, between, say, 16,000 years ago and 5,000 years ago there was a revolution in agriculture. Different plants and animals were domesticated around the globe depending on what native flora and fauna were available but almost all people, all around the world at almost the same time, started to change their lifestyle from nomadic hunter-gathering to settled farming. This is remarkable. It is almost as if Man was pre-programmed for this change.

There is a popular misconception that life as a caveman was hard work – chasing game, making tools, searching for food must have taken up all the time. In fact, life as a hunter-gatherer is relatively easy. Studies of contemporary hunter-gatherer societies indicate that only a few hours a day are spent actually gathering food and the rest is pure leisure time. With the more abundant resources available to our ancestors, they were probably living the good life. Settled life is much harder. Settlement allows an increase in the number of the group, specifically in non-productive members of the group – the very young and the very old – but these are mouths to be fed which increases pressure on the productive members. Once you are settled you may have to go further to find food, especially after you have killed all the local animals and eaten all the local vegetables. This leads to farming and, as any farmer will tell you, farming is hard work. So why settle down?

It is easy to sit here and see disadvantages in settlement over hunter-gathering but it is much more difficult to predict these problems beforehand. If early humans settled in areas where game was plentiful and there was plenty of edible local flora they probably thought these conditions would last forever. They probably did not even consider their long-term impact on the environment. They probably did not foresee that they would use up all their local resources in the same way that we fail to foresee that we will use up all the fossil fuels. They probably did not plan their population growth any more than we are planning ours today. Life just happened.

Then one day the game was gone, the trees were bare, the roots and berries were eaten. What was Man to do? Most animals, if the food runs out, will either move somewhere where food is more plentiful or will starve. The problem with moving is that Man's population, even by this time, was such that there were few viable free areas, that is liveable areas unoccupied by other people, to move to. With his relentless expansion Man had by now taken over the whole habitable globe.

But, of course, this crisis was not the end of Man it was the making of him. They say that necessity is the mother of invention and the invention this crisis precipitated was agriculture. Man neither moved nor starved – he simply grew more food where he wanted it. The significance of this cannot be overstated. This is the basis of all of civilisation from the very earliest to our current, complex societies.

There has been speculation as to whether people settled in permanent sites and were then forced by population pressure and depletion of resources to develop agriculture or whether they reaped and planted as part of their seasonal migrations and then eventually realised they could grow adequate food in one place and so settled there. As with most of these debates there is probably some truth on both sides but the first scenario seems more generally likely. There are still nomadic herders living in the world, driving or following their animals as they forage for food but these are few and far between. Evidence for settlement precedes evidence for agriculture in most cases.

Evidence of gathering, which is the precursor of agriculture and of early agriculture itself, is harder to come by than evidence of hunting. Soft seeds, roots and fruits are not as well preserved in the fossil record as hard bones and teeth. If the remains of a dog or a goat or even seeds are found in association with human remains it is very hard to establish whether this implies domestication or not. Furthermore, fossil evidence is best preserved in an arid environment rather than in moister areas where agriculture may be more likely. For these reasons, it is impossible to say exactly where and when agriculture started. The first hard evidence we have comes from the Middle East and consists of flint sickles and grinding stones dated about 14,000 years ago. These comparatively sophisticated tools imply that agriculture was fairly well developed by this time and may imply that the first glimmerings of agriculture occurred much earlier.

There are many theories as to how agriculture began but it was quite probably as simple as noticing that where seeds dropped, new plants grew and then exploiting this observation. Of course, while we may call this simple, it is something no large animal had ever done before.

I say 'large animal' because ants have been growing fungus gardens and 'farming' aphids for millennia. Whether one would call this 'agriculture' is debatable: the ant/fungus and ant/aphid relationships could be viewed as a symbiotic as both sides benefit whereas agriculture is the exploitation of one species for the exclusive gain of another.

The first plants to be domesticated were grasses and they are still the most important food crops in the world today. Grasses had been around for a long time. The best grasses from Man's point of view were the cereals which have relatively large, nourishing seeds grouped in a seed head. Man, being omnivorous and being curious, probably ate anything that came into his ken. This would have included wild grasses. Man probably started harvesting wild grain well before he ever started growing his own. Domestic grasses today include not only the cereals we eat but also the grass on our lawns and playing fields and many species of bamboo. The first domesticated grasses were wheats which were brought under human control in the Middle East about 12,000 years ago.

Before continuing it is worth considering what is meant by domestication. Domestication means that Man controls breeding. You may argue that your cat breeds whenever and wherever it can and this is not actually under your control, but it is. You can spay your cat or drown its kittens. It is domesticated. The original wild grasses are pretty poor compared to the high-yielding strains we grow today – the grains were smaller and fewer, they broke off the stem more easily, they had heavier and more clinging husks – but this soon improved. In fact in most cultures where agriculture has started there has been a rapid improvement in the quality of the plants grown. For example, the first domestic wheat (einkorn wheat) cross-bred with other wild grassed to produce emmer wheat which has larger seed heads. This in turn cross-bred with other grasses to produce the bread wheats we eat today. These cross breeds were stronger, higher yielding, less brittle and had lighter husks than their wild forebears.

Exactly how Man achieved this result at so early a stage of his agricultural adventure is a moot point. There is obviously some selective breeding going on but whether this was a deliberate, directed attempt to improve the strain or the happy outcome of chance events is impossible to say. One suggestion is that the very best of the crop was strewn over the growing fields as an offering to the gods and that this continual re-sowing of the best plants led to gradual improvement. When I say improvement, I mean an improvement from Man's perspective and not the plant's. The wild grasses had done perfectly well before Man's interference. They had been evolving for about 25,000,000 years and had developed their low yield, brittle stems and heavy husks under the normal rules of natural selection. Man is the one who increased their yield so that they might 'increase their numbers'. Prior to Man's interference they had survived quite adequately without this.

This improvement of the strain has continued throughout Man's history. As with almost all trends there is a slow start, a steady and gradual build up, a few leaps forward and suddenly, in the last century or so, a huge increase in pace. The cultivation of wheat spread from the Middle East to India by about 6,000 years ago and to China by about 2,000 years ago. Wheat cultivation continued much as ever with gradual improvements in the machinery used and slow improvement in the plants until the Agricultural Revolution about 400 years ago. By rotating crops and deliberate, selective breeding, higher yields became possible. In the last one hundred years or so, as we have learnt more of genetics, new hybrid strains have been developed. These are the highest yielding, biggest grained, best in all things wheats ever grown. Unfortunately they are also infertile. Now we are playing with GM (genetically modified) strains. We can build wheats for any soil conditions – arid, wet, acid, alkaline – resistant to pests and diseases, super-high yielding, everything you could ask for in wheat and more.

However there is a downside. In 1970, 75% of all the corn grown in the US (and that is a huge amount of corn) was of a single hybrid variety. If a specific blight had attacked that strain there could have been disaster. Which brings up another pertinent point: as Man has developed cereals, so the blights and rusts which attack those cereals have also developed. No sooner does a rust-resistant strain appear than a new rust appears to attack it.

At about the same time as the domestication of field crops there are the beginnings of animal domestication. Probably the first animals to be domesticated were dogs, evidence for which comes from Israeli and Iraqi sites about 12,000 years ago and from America about 10,000 years ago. In America at least these were probably bred for food. Dogs were followed closely by sheep and goats by about 9,500 years ago. There is evidence of domesticated cattle in Greece about 8,300 and in India about 7,000 years ago although it is unclear whether the two are related or whether wild animals were domesticated independently in both areas. In Europe and Asia horses, pigs and chickens were domesticated while in the New World llamas, ducks, guinea pigs and alpacas were taken into human custody. There are many species now under Man's control including insects (honey bee, silk moth), birds (chicken, duck, turkey) and fish (carp, trout).

The early domestication of the dog, Man's best friend, probably started with keeping wild wolf cubs. Wolves are pack animals, used to having a dominant male who calls the shots. If cubs are taken young and isolated from their peers then Man can become an ersatz pack leader. Although dogs were probably initially domesticated for food, they have other uses – their natural territoriality can be used to make them guard dogs; their hunting skills can be harnessed to Man's needs; their warm coats can be a living blanket; their relative intelligence makes them lively companions.

By selective breeding Man has transformed his dogs into tools for his own use. For hunting he has produced hounds, pointers, setters, retrievers and specialist killers like dachshunds (for badgers), fox terriers (for foxes) and wolf hounds (for wolves). He has guard dogs, sheep dogs, cattle dogs. The Inuit have huskies for their sleds. There are guide dogs for blind people, companion dogs for the mentally ill, toy dogs for people who like toys and plain mutts for people who just want a pet. Greyhounds and whippets are raced for Man's sport and other dogs are made to fight. Dogs are still eaten in some countries although there is a taboo against this in others. Domestic dogs now outnumber wild dogs by many millions.

Domestication of goats, sheep and cattle may have come about when young animals whose parents had been killed were taken in by the group. There have been huge improvements (from a human point of view) in these animals. Cattle probably came from the wild aurochs, the last of which were killed in Poland in 1643. By selecting wild animals which were docile, fat and stupid and then carefully cross-breeding these and eliminating any animals which do not accord with his plan, Man has produced animals which are really docile, really fat and really stupid – evolution by survival of the least fit.

Now, we farm in crowded feedlots and batteries with animals tailored to those conditions. We now have specialist milk cows that produce so much milk they would burst if not relieved of it, pigs that do nothing but get fat, chickens that do nothing but sit and lay eggs which will never hatch. However many of these traits are only superficial. Feral animals soon revert to type – feral pigs quickly turn from fat and lazy pork balls into aggressive razorbacks, feral dogs revert to their hunting past, fancy pigeons become flying rats.

Many domestic animals have multiple uses, for example cattle – their hides make good leather, their dung can be used to fertilise the fields, their strength can be used to pull a plough, their bones and horns can be used for tools and artefacts. Chickens can be used for meat or for eggs or their entrails can be used to divine the future. Horses can be eaten, their tail and mane hair can be used for thread or for stuffing, their hide makes good leather and they can be ridden or used to carry heavy loads or to pull a cart.

An exception is the cat. The cat has been domesticated but, while the relationship is still one of exploitation, it is not quite so clear who is exploiting who. Cats were worshipped in ancient Egypt and they are still revered by some today. Domestic cats can and do catch birds or mice or other small animals but as rat-catchers we have terriers and for hunting birds we have other dogs. Most cats' hunting skills are only used when they go feral.

It is hard to say what Man gets from keeping cats or pets such as parrots or hamsters other than a perceived relationship with another animal. Man is the only animal to keep pets. The concept is totally alien to the natural world.

As communities settled down so there arose new needs and new inventions to supply those needs. At much the same time that Man was beginning agriculture, he also developed pottery, spun thread and woven cloth. The first pots were made by coiling ropes of clay into the desired shape but soon came the wheel. In Europe the first evidence of wheels is on wheeled carts about 4,000 years ago and then potters' wheels about 500 years later whereas in Egypt the potters were using wheels about 4,500 years ago but the first wheeled carts date from 1,000 years later.

Also at about this time Man discovered metal. The first metals to be used were copper and lead, both of which occur naturally in relatively pure form and can be easily worked. They were being worked in southern Iraq by about 8,000 years ago. The problem with copper is that it is quite soft and so Man made it harder by adding a little tin to make bronze. This was in use by about 5,000 years ago. Bronze enabled Man to make much more efficient axes to cut down the forests and much more efficient arrows and spears to kill with. The use of bronze made copper and tin valuable resources to be exploited.

These are only examples of a whole swag of new inventions which came about in the period between the Upper Palaeolithic revolution about 30,000 years ago and the development of cities about 6,000 years ago. Many of the most fundamental inventions span continents and cultures which could not have been in even indirect contact – for example, agriculture occurred in the Middle East, in the New Guinea highlands and in Mexico; spear throwers were used by both Australian Aborigines and Canadian Inuit. Bows and arrows, spinning thread and weaving cloth, the use of the wheel in both transport and pottery, working metal, the

111

development of cites – all these and many more occurred in geographically separated regions. Traits like body-piercing, tattooing and the use of gold and precious stones for jewellery take many different forms but are common to many unrelated cultures. Almost all cultures have some form of religion. Almost all have some form of art. All have spoken language and most have written language. Most have a calendar of some sort. Almost all have music although instruments and tonalities vary hugely. Almost all have alcohol or other drugs.

Humans are the only animals which mutilate themselves and it seems to be a common trait in many cultures. Which body-part is mutilated and how is a matter of taste – for some it is removal of a finger joint, for others a tooth; others stretch the neck or the earlobe or squash the head or the foot; for many others the targets are the ears or the penis and probably the most common of all is simple scarification or tattooing. Why humans cause themselves injury, risk infection and septicaemia and in many cases suffer considerable pain in this way is an enigma. It is something that started very early – there is evidence of self-mutilation in some very early finds. This seems to be another example of separation from nature but this time a deliberate attempt to defy and deny nature, to become to some extent Man-made.

The problem with settlement based on growing crops and tending herds is that you need somewhere to grow and tend. Inevitably agriculture, even ancient agriculture, has an environmental impact and the long-term effect of this impact is negative for all other species except Man and his domesticates. As settlement increased and numbers grew Man needed to create open fields for his crops and so began the forest clearing which is still going on today. To protect his crops and his herds he needed to eliminate vermin and predators – lions and elephants used to roam the Middle East up to biblical times but do not do so anymore; wolves and bears were once common in most of Europe but now only survive in isolated pockets. To grow crops you also need plenty of water. In temperate Europe this is not too much of a problem but in the Middle East where rain is less reliable this requires irrigation. Quite apart from the immediate physical impact of irrigation ditches and canals and the impact of reduced water flow downstream, irrigation gradually builds up salts in the soil being irrigated until it eventually becomes too saline to use.

Of course this was a very gradual process. I am not suggesting that the environmental impact of the first farmers was anywhere near the scale of destruction going on today but there was undoubtedly some impact. Some of the first settlers probably used the same slash-and-burn technique still being used today. In this form of agriculture a section of forest is deliberately burned and cleared. The ash from the fires creates a fertile environment for a short while but once this superficial goodness has been used up by cropping, the soil is often too poor to support intensive farming and so the settlers choose a new patch of forest to slash and burn and the cycle continues. Slash and burn agriculture is very destructive but the forest will regenerate in the burnt areas after a few decades or centuries. As long as there is a big enough forest and a small enough population then the forest has time to recover before it comes under attack again. In ancient times there was plenty of forest and few people so no problem. Unfortunately the areas under attack today have diminishing forests and increasing populations so the system is not sustainable.

The environmental impact of the first farmers would have been noticeable but relatively minor. This is not because Man at this early stage was much closer to nature or had more environmental awareness, concern or consideration than he has today, but simply because the initial populations were small. What has changed is not so much what is being done as the number of people doing it. Most of the environmental problems around today caused by agriculture – salinisation, soil erosion, forest clearing, problems with feral and introduced animals – are ancient problems. The change from wheat to more salt-tolerant barley in ancient Middle

East agriculture is quite probably due to the increased salinity brought about by irrigation.

Of course if the soil gets salty or becomes a dust bowl then with plenty of space and a small population you just move on and do it all again. And so Man has, over the centuries, expanded his farming culture throughout the globe and with him he has carried his domesticates and his parasites.

Agriculture was not the first or only reason for environmental attack by Man. Man already had more reasons for killing than any other animal – for food, for clothing, for bones and horns for tools, for self-protection, to eliminate competition and quite possibly, even at an early stage, for bravado and for sport. Agriculture just provided a few more reasons.

Once agriculture started it became a runaway success. Agriculture tended to create a surplus of food but instead of settling back and enjoying the easy life, numbers soon grew to use up the surplus. There are few checks on human population and with a ready food supply the stage was set for the beginnings of the population explosion which is still continuing today.

Settlement sizes began to grow. Nomadic hunter-gatherers need a large area to support a small population and so tend to be widely dispersed. In contrast, farming leads to concentration of population in the fertile areas suitable for growing. The most fertile areas about 10,000 years ago were river valleys such as the area between the Tigris and Euphrates rivers known as Mesopotamia in what is modern Iraq. These areas were well settled by about 6,000 years ago. Villages with populations in the hundreds were common and the larger settlements covered about ten hectares and contained about 2,000 people but within a thousand years things got much bigger. By about 5,000 years ago there were real cities – Uruk, Kish, Ur and others in Mesopotamia, Jericho in modern Palestine, Catal Huyuk in modern Turkey. Populations varied but the population of Uruk 5,100 years ago is estimated at 10,000 and only 500 years later this had grown to 40-50,000 in an area sprawling over 100 hectares. This by any measure is a considerable size and there were several other cities this big.

It is barely credible that in just a few thousand years, Man had moved from chipping flint to casting bronze, from hunting animals to farming wheat, from travelling from cave to cave in a small family-based group to living in brick houses in city communities numbering tens of thousands. In Mesopotamia by 5,000 years ago and in many other places at around the same time or within a few thousand years, there were all the elements of modern society. There were rulers and the ruled, rich and poor, merchants, accountants, tradesmen, artisans, farmers, bureaucrats and labourers. There were large cities made of brick with public buildings, private houses, temples and palaces. There were written records, calendars, laws and the beginnings of money.

This is remarkable enough but the fact is that very similar developments were taking place all around the globe at about the same time or slightly later, seemingly independently. In China's Huan He province, in the Indus and Ganges valleys of India and in Central and South America farming settlements were growing into cities with all the same structures as the Middle East. The crops farmed, the animals domesticated and the farming techniques differed but in all cases the land was 'improved' by irrigation, draining or damming, fields were enclosed, wild lands were tamed. The same sort of improvements through selective breeding applied to most domesticated animals and crops. We also see brick buildings, pottery, metal working, spun thread and woven cloth used in most areas. There are musical instruments, evidence of organised religion, kings and peasants all over the world.

Most of the cities of this time have large buildings which are taken to be religious centres. Cattle seem to have been popular objects of veneration, certainly in Catal Huyuk and in the Indian centres and later in the Minoan civilisation of Crete. In these early cities we also get the first pyramids. Pyramids are usually associated with Egypt but there were ziggurats in Mesopotamia many years before and in Central America some years after and the same pyramidal form is seen in India and the Far East.

Another striking fact is that all of these cities show evidence of rich people and poor people. There are large, opulent, well-decorated palaces of the rich and there are the simple houses of the ordinary people. Even the poorest Man's house is better than any animal's, but the largest houses show the wasteful extravagance and conspicuous consumption which has always typified the rich and powerful. Among the very rich there is and always has been a tendency to want to leave the biggest mark on the world – to have the biggest palace, the highest tower, the most splendid tomb. Or perhaps it is that those who wish to leave a mark on the world are driven to become the most successful. There never seems to be a point where enough is enough.

These were proper cities with communal areas and private areas, palaces and slums. The logistics of food supply, sanitation, law and order and public works were well beyond the level of individual solution. For any settlement to grow this big there needs to be a level of organisation far greater that that required for simple cooperative hunting and gathering. Major works such as building irrigation channels would have required planning and management for the good of the whole community. This implies a planner or planners.

It is most likely that the first settlements were clan or tribe based and that decisions affecting the whole settlement would have been made by the leaders of the clan. As settlements grew and clans mixed, the clan would have become subsidiary to the community and new forms of governance would have been invented. There is no direct evidence of the political organisation of the first settlements – they may have been governed by a king or a council or a combination of both or perhaps some were one way and some another.

The obvious candidates for planners and managers are those who guided the people in spiritual matters – the priests – and these were the first community leaders. We can be certain about this because the next great leap in organisation was made at around this time – the invention of writing.

The first written records we have are from Mesopotamia and date from about 5,000 years ago. These appear to be tallies of goods – perhaps commercial bills or tax returns. It is generally thought that writing began as pictures and signs – a picture of an ox means an ox, a picture of an ox with three other marks means three oxen. These pictures became more and more simplified over time, gradually changing from actual pictures to mere symbols. Eventually writing developed into a code where each symbol represented a sound rather than an object – the symbol for a bee for example represented the sound 'bee' rather than a winged insect. Using this method it is possible to squeeze more and more information into fewer and fewer characters until we reach the 20 to 30 character alphabets of many languages today. Writing is another global invention and it goes hand in hand with another invention of about this time – the calendar. Writing and calendars, although of vastly different designs, were both also invented in the Americas and the Far East. It is probable that cites beyond a certain size are simply not possible without some form of written records.

While spoken language is universal, there are many cultures which have never developed writing. Writing is not easily picked up like spoken language, it must be deliberately learned, and there are many people who cannot read or write although they can speak perfectly well. This difficulty means that written language is for the elite, the rulers, while the peasants have no need or use for it.

With the invention of writing we leave pre-history and move to history proper. The stage is set for Man's next big invention – war.

Man versus Wolf

It is interesting to contrast the fortunes of Man with another contemporary species, for example the wolf.

The Dire Wolf of the last Ice Age was comparable in size and weight to Man. This went extinct along with the rest of the megafauna about 7,000 years ago. The most common wolf now living is the Grey Wolf (*Canis lupus*) which ranges from about 35 kilos up to about 55 kilos although exceptional specimens up to 79 kilos have been recorded. There are also the Red Wolf, the Ethiopian Wolf and the Maned Wolf although the latter is not a true wolf. Most species are endangered.

Wolves evolved in the mid-Pliocene about 5,000,000 years ago, well before Man. Not unlike early Man, they live in family-related groups. They have some level of communication using facial expressions, vocalisations, scent markings, body language and behaviour. They hunt as a group and they share the spoils of the kill. They are at the top of their food chain with no natural enemies except Man. It is impossible to reliably estimate animal populations prior to a couple of hundred years ago – before that time nobody really cared – but we do know that wolves were for many millennia the most widely distributed large mammal in the world ranging throughout Eurasia and the Americas.

Wolves have always been competitors to Man. There are many documented instances of attacks on people (although these may have been by lone, rabid animals) and many attacks on flocks and herds. Because of this, wolves were eradicated mercilessly – the last wolf in Great Britain for instance was killed in 1743. In the rest of Europe, Asia and America the wolf has not been totally exterminated but has been reduced to a tiny fraction of its original population. Wolves have now been confined to the unwanted places of the Earth. There are now several sub-species possibly due to the fact that Man has created geographically isolated populations. Wolves are still hunted in some areas; in some areas they are protected and in some areas they are even being re-introduced into the wild.

In a one on one situation it is hard to say who would prevail: Man or wolf. The wolf certainly has the upper hand in killing equipment with its fangs. It is hard to see how a Man could kill a wolf bare-handed, but Man has his cunning and that should never be underestimated. However, wolves rarely fight one on one. Their tactic is to chase with the whole pack and then move in for the kill. If a Man were to face a wolf pack alone and unaided his chances of survival would be slim. The wolves could outrun him and outlast him. While one Man might fight off a single wolf, a pack upon him would be his end. In the 'natural' scheme of things, the wolf would rule.

But Men do not come in ones, they come in hordes. And they are not unaided; they have their weapons, their traps and their poisons. So the wolf does not prevail. From being one of the most widely distributed mammals on the planet they now hang on as remnant populations in isolated communities. On the other hand the domestic dog (*Canis familiaris*) which has associated with Man has gone from strength to strength.

The wolf was once Man's competitor; now it is something to be pitied and protected. The same story could be told for the bear, the lion or any of a hundred other species.

This is Man's success.

And so on…

We are now entering the modern history of the world – the final milliseconds of our world evolutionary day. This book began with the Big Bang 20,000,000,000 or so years ago. We are now in the last 10,000 years that is the last 0.00005% of time so far. Time stretches on ahead, possibly without limit. To spend a whole chapter on this particular 10,000 years, a mere tick in the grand scheme of things may seem extravagant and it is. However, much has happened in these recent years and it has shaped the world as we see it now, so it merits its own space.

On the evolutionary theme, there have been a few changes in the last ten thousand years. In general Man has become taller, stronger and longer-lived. To a great extent this is due to improved nutrition and improved health care but that does not really explain the increase in average height which could be as a result of cultural selection. Also some people have changed colour from black to white and several shades in between with concomitant changes to hair colour, hair texture and eye colour. These changes are taken to be a result of natural selection for sun protection but could just as easily also be cultural selection. These superficial changes have been assigned great significance in some quarters because they are so obvious and because Man as a species seems obsessed with superficial difference. Apart from these minor changes there have been no significant physical developments – teeth, stance, facial features and brain have hardly changed in the last 50,000 years. A

Cro-Magnon from the early Palaeolithic, suitably dressed, would not stand out in, say, a Saturday afternoon football crowd in a modern city.

Any history of the world since Man rose to dominance must in fact be a history of Man. It is inconceivable that one should write a modern history of a species other than Man. A history of, say, horses, over the last few thousand years would make pretty dull reading – horses simply have not done that much in that time or in the last few million years for that matter. The major changes for horses in modern times are because of Man. New breeds like cart horses, pit-ponies, pacers and quarter-horses; the re-introduction of horses to the Americas and elsewhere; the virtual extermination of the original wild stock: horses themselves had little to do with any of these other than being passive subjects of Man's exploitation. Most animals just go about their lives – they are born, they live and they die the same as their parents did and their parents before them and their parents back through any number of generations. This is definitely not the case with Man.

From the point of view of, say, a bear, the last 10,000 years has not been a golden age. As a bear might see it, Man spread around the world like a plague, hunting and killing, invading and clearing the forests. As he spread he formed into large tribes. These tribes claimed a slice of the world as their own to use, and squabbled over boundaries. Tribes split and borders changed again and again. Through it all Man's numbers grew. He went from sticks and stones to spears to bows to guns; from stone axes to bronze axes to iron axes to chainsaws. His traps and poisons improved. His roads and walls and fences cut through the wild lands. His villages grew into towns and his towns grew into cities and his impact on the world increased. Meanwhile the quarrels continued. The reasons for the squabbles changed but the squabbling went on as ever. And Man kept on increasing and taking more land and killing more bears. And the last few hundred years have been the worst of all. To a bear it would not make much difference whether it was

Assyrians fighting Hittites or The Allies fighting The Axis – it's just Man, fighting again and bad news to everybody.

What about the art, the culture? Many people consider art and culture to be the true measure of Man. His elevated sense of aesthetics, his philosophy, his technology to them demonstrate Man's higher nature, but the bear would probably have a less rosy view. It is unlikely that a bear could tell the difference between a cave painting and a Picasso or would care if it could. Man's technology would be nothing but an imposition. Everything Man has ever done is for the benefit of Man. Even if we now pretend to manage the environment or protect endangered species it is only to ease our guilty conscience or to preserve a curiosity for future generations.

The modern history of the world is a story of conflicting and complementary trends. There is a trend towards increasing size and complexity of social organisations, increasing scientific knowledge, increasing sophistication in art, literature and general culture. At the same time there is the trend towards confrontation, conflict and war. Going along with both is the trend of continuing, unquestioning exploitation of resources. Above all else is the overwhelming need for more – more of everything. This has been so from the very first and still continues today.

With the development of agriculture and villages and towns we begin to have a new concept: ownership of land. Up to the time of settlement or shortly before, land had just been land. But once people started working their fields, they gradually became *their* fields. Now land was owned by one or by another. Once land is owned it can be bought, sold, traded or stolen or merged with other lands until one person may end up owning a substantial slice of the world. Land and other possessions can also be passed on from parent to child. Inheritance of this sort is an entirely human trait; it does not happen in the natural world – territories are not passed down through generations – there are no animal dynasties. Perhaps for a generation or two the progeny of the original dominant animals may rule the roost but only on their merits if they are more able than their rivals. In most cases once the young reach a certain age they are expelled from the parents' territory to fend for themselves.

Human land ownership is quite a different thing from an animal having a home territory. No other animal, except for Man, lays claim to the whole resources of a whole territory to the exclusion of all others, for him to own and dispose of as he pleases. Not all Men make such claims – the Aboriginal peoples of Australia regarded themselves as owned by rather than owning the land; some of the Aboriginal peoples of America regarded themselves as custodians of the land to hold in trust for generations to come.

Be that as it may, it is this concept of ownership of land that is at the heart of human history. As well as the continual struggle over who actually owns the land there is the other side of ownership – the ruthless exploitation of the resources of the land – the timber, the minerals, the game, the soil. And with this, as with everything else is the need for more: more land, bigger empires, new colonies.

The influence of the early cities expanded into the surrounding villages as trade increased and more land was brought under control. As Man's population grew and town sizes increased so did the amount of land under cultivation. Agricultural techniques did not change much – the same sort of ox plough used in early Sumer was still in use until about a thousand years ago – so the only way to produce more food was to farm more land. This was not a fast process but it was an inexorable one. The cities grew into city states. The states expanded into empires. People began to see themselves not just as individuals but as being from one country or another. Rivalries began between empires and these soon escalated into open conflicts. One empire gained dominance and then another. Most of these early conflicts were about resources – land, riches, slaves, etc. This is the 'struggle for existence' that Malthus was so concerned about – it has been going on since the very first settlements and it still continues today. The very first cities had massive walls and battlements designed to resist attack.

In evolutionary terms, war is nonsense. Any animal which wilfully destroys itself must surely be an evolutionary loser and you might expect that such behaviour could not exist in a system of evolution by survival of the fittest. Of course it does not. War does not exist in the natural world.

No animal makes war like Man. The obvious reason is that few other species have the spare capacity to destroy thousands or millions of viable individuals and those that do, octopi, for example which may produce millions of young, seem to do so for the benefit of others rather than as a futile act of self-immolation. Man not only has the capacity, he has the will and the means and he has been killing himself from his very early beginnings. There have been many analyses of war and violence, searching for reasons hidden in Man's animal heritage but they are unconvincing. When animals of the same species confront one another, they generally do so as individuals and generally for some immediate, specific reason such as mating rights or possession of a kill. What's more it is often possible to establish the winner and loser by comparing size or physical attributes without coming to blows at all. Thirdly, even when animals do actually get into a physical struggle the loser is usually quick to call it a day before he is badly hurt and fourthly the winner usually lets the loser go with a whole skin instead of pressing the advantage to the ultimate and killing him.

This is not so of Man. Wars are not between individuals but between abstract concepts like countries and empires. They are often fought for a number of reasons which have been simmering sometimes for years and very often there are underlying causes which are not even apparent. Winners and losers are harder to predict. A small, well trained, well equipped and strongly

126

motivated force can often overcome much larger forces and then there are techniques of guerrilla war, sabotage and internal conflict whereby a small minority can inflict a great deal of damage even when overwhelmingly outnumbered. Often the eventual inevitable losers will not surrender early but will fight to the last individual or will fall back to continue their struggle by stealth. And Man is not slow or hesitant to punish the losers to the extreme.

Man's war, like so much else about Man, is his own invention. Man is no more 'naturally' warlike than he is 'naturally' agricultural or he 'naturally' wears clothes. Man has chosen to pursue war as he has chosen to pursue agriculture and has pursued it with a thoroughness and invention that are frightening. The technology of modern war and of current weapons is among the most advanced technology applied in any area. The billions upon billions of dollars spent on military adventure in the hope of finding ever better ways to kill more people far outstrips money spent on agricultural research to assist famine relief and keep people alive. Declared and undeclared wars are prevalent in many areas of the world along with terrorism and general thuggery and have been throughout history.

When one considers the number of people killed in such conflicts and considers that these are for the most part young, potent males then one begins to see that human population could be much, much greater. Wars have been brought about in many cases by population pressure and they are to an extent both the cause and a rather drastic cure.

But Man's inhumanity extends beyond war. In most of human history, the inhabitants of invaded countries have either been treated as a resource just as much as the timber or minerals or they have been treated as an inconvenience to be pushed aside or killed as necessary to clear the way for the invading peoples' ambitions. They have rarely been treated as people and even more rarely as equals to the invaders. Slavery has been practiced in many cultures openly or covertly and as slaves people have been treated worse than animals; transported in conditions worse than cattle; worked literally to death. Even worse is ethnic cleansing, genocide and torture: the very words are a disgrace to humanity but they all exist or have existed. You will search in vain for any such behaviour in the natural world.

All over the world, in Europe, Asia, Africa and the Americas, empires arose, dominated their particular corner of the globe for a while and then crumbled as the next invaders took over – Egyptians, Persians and Romans had a go, among others. In the East, in the Indus valley and China, similar stories can be told. In the Americas there were Mayans, Olmecs, Incans and more. Grand cities were built, sacked and re-built. The major religions were formalised, providing more on-going reasons to fight. Ocean-going ships were built and the last remaining natural sanctuaries – the far-flung Pacific islands – were invaded. Trade in goods and ideas became world-wide along with the unforeseen and unwanted concomitants of trade and travel – transport of feral animals, pests and diseases. Countries were invented. Lines were drawn on maps and then rubbed out and redrawn. Kings and queens came and went, battled, won and lost and the modern world emerged. Settlement burgeoned everywhere. As open space was found so it was filled, and so more space was opened. In Europe, the forests were cleared for building houses and for building great fleets for trading, fishing and fighting. The East was plundered for its silks and spices, Africa for its gold, ivory and slaves. As Europe started getting crowded so the Old World invaded the New, drove off the native inhabitants and began the process anew.

Then, about three or so hundred years ago, there was a sudden surge of innovation in science and technology in all fields which is still going on today. For the sake of brevity I will refer to this as the Technological Revolution. I include not only the early innovations to industry brought about by the use of steam power and mechanisation but also include the advances of the nineteenth and twentieth centuries in all fields.

The history of humans as a species can be split into several extremely unequal periods: the emergence of modern *sapiens;* the exodus from Africa and invasion of the rest of the world; the beginnings of agriculture and permanent settlement; the ten or so thousand years from then up to the Technological Revolution, and the few hundred years since then.

Although technology improved greatly between the time of the first settlements and the Technological Revolution, the fundamentals remained the same. Up to a few short hundred years ago, everything was on a human scale and was powered by the effort of humans or animals or in a few cases by wind or water. Although eighteenth-century ships were much more sophisticated than the earliest vessels they were still made of wood and powered by the wind. The fastest means of transport generally available was still horsepower. The best lighting for most people was candles. Most people lived and died within walking distance of their birth place.

The rate of change from, say, 10,000 years ago to about 300 years ago is nothing compared to the massive acceleration since then. In a miniscule amount of time, not even measurable on a geological timescale, Man has mechanised and industrialised the world. If you consider for a moment the innovations of the last few hundred years – the use of electricity, the internal combustion engine, air travel, advances in communications, computers, etc. – and consider the impact that these have had on the average man-in-the-street it becomes apparent that the last 300 years have changed life as we live it more than the previous 10,000 years. When Thomas Edison patented his electric light bulb in 1879 he usurped the night and transformed the world. The use of the motor-car as an everyday conveyance has completely changed how we consider distance. The telephone has become an indispensable accessory –

we are at all times in reach of instant and constant communication. There is hardly any field of human endeavour which has not been fundamentally affected.

While an Iron Age peasant might be amazed by the yield produced from seventeenth-century farming methods, he would at least be able to relate to the farm implements and the horses pulling them. Contrast this to the mono-culture deserts of modern agriculture with their vast uninterrupted seas of GM wheat and monstrous tractors towing intricate multi-processing machines. Similarly, a citizen of ancient Ur may have found the crowded streets of seventeenth-century London to be a slightly alien environment but they would seem almost familiar compared to the concrete urban sprawl of a twentieth-century city like Los Angeles with its neon-lit, smog-polluted centre and endless car-filled freeways.

Man as a species has always been at odds with nature but now we seem to be positively attacking the world. We are building, fishing and farming on an unprecedented scale. As our ancestors once tunnelled into hills to pick at copper veins with stone tools, now we level mountains to strip-mine uranium; as we once slashed and burned our way into European woodlands to eke out a living, now we clear-fell Amazon rainforests. We are changing the very contours of the world by building off-shore islands, drowning valleys to create massive lakes for water storage and hydro-electricity and stripping down mountains to pillage their mineral wealth. We have technology which allows us to go anywhere in this world or out of it and do anything we please when we get there. There is hardly a corner of the world so wild or so inhospitable that Man has not now invaded it and turned it to his use.

In most human population graphs the graph looks fairly steady up to about a couple of hundred years ago where it suddenly zooms towards the top of the page. This sudden leap disguises the fact that even if the graph were not to include the last few hundred years you would see a slow but steady rise. There is a noticeable dip in the thirteenth century or so when the Black Death took a good chunk of the world population but apart from that the line goes up. In contrast to this the population of almost all other large animals (apart from Man's domesticates) has been steady or in decline over the same period. The recent massive acceleration is due mainly to the surpluses produced by the Technological Revolution plus improvements in hygiene, diet and medicine. Inventions like penicillin, antibiotics and inoculation have reduced the death rate considerably. However to an extent it is an inevitable consequence of the previous build up – we achieved critical mass and exploded.

Man has had a large impact on the world for two reasons: firstly Man treads heavily on the world and secondly there are an awful lot of us. In most areas where Man has settled, he has thrived. Man's population has increased steadily despite the ravages of famine, flood, pestilence and war. About 10,000 years ago, when the first cities begin to appear, the population of the whole world was perhaps 5,000,000. As agriculture developed and more land was cultivated, so the population began to grow. The size of cities, states and empires was ever increasing. The Roman Empire at its height included all of Europe south of a line from Holland to the Black Sea plus bits and pieces of Asia Minor and North Africa. There were 50,000,000 people in 5,000 administrative divisions. Rome itself housed 1,000,000 people. In China the Qin Empire was even bigger at about 60,000,000 people. In Mexico, the city of Teotihuacán was the sixth largest in the world with 200,000 inhabitants. All in all at the height of the Roman Empire about 1,800 years ago, world population was only somewhere around 200,000,000.

There have been periods of growth, periods of stasis and periods of decline but in general the human population graph shows an inexorable growth, even if we ignore the extreme boom of the last few hundred years. This applies world-wide, in all communities at all times – all cities tend to grow. Man tends to expand to fill the available space, and Man decides what is 'available space'. Contrast this with nature. A leopard may have its territory and may defend that territory but it does not try to expand its territory at the expense of others simply so that it can be the leopard with the biggest territory. Animal populations are regulated by nature and cannot grow out of proportion to it – if there are too many mouths to feed it means that some will starve; there is no option of growing more food. In a predator/prey relationship, populations may boom and bust in cycles but they stay in relation to each other and to the rest of their environment. Man, being the master of nature, can do as he pleases.

Man has now changed the world so radically that it is hard to say what is 'natural'. By hunting, by settlement, by changing the landscape, Man has so changed the rules that if you look at the typical countryside of England for example there is scarcely a natural thing in it. The land is parcelled into fields growing crops and animals modified by Man. There are roads and walls and houses, bridges and canals. And within this dwell not only the natural denizens – the wrens and finches, the foxes and badgers – but also exotic imports and feral escapees. Even such 'wilderness' areas as the Scottish Highlands have been hunted for millennia: there are human tracks in the most remote of areas. One of Darwin's examples of natural harmony is the English hedgerow – his famous 'tangled bank'. This environment did not exist to any great extent until the enclosures of the Middle Ages – hardly 'natural'.

Elephants were once the absolute kings of their domain, the top of their food chain. Not even a lion will normally take on an adult, healthy elephant. In the natural scheme of things, what will inevitably kill an elephant, if no other accident befalls it, is its teeth: after its last set of teeth wears out the elephant cannot chew and will die of malnutrition. Now elephants run from Man. They shun him. They live in man-made enclaves where they are still hunted. Can their behaviour be called natural?

Until about 5,000 years ago elephants were common throughout Africa from the Mediterranean coast to the Cape. When the Sahara dried up the population was split into the northern herds which ranged from Tunisia to Morocco and beyond and the others south of the Sahel.

Elephants and before them, mammoths, have always been hunted for ivory – some of the very first recognised art works are carved ivory figurines. Ivory has long been a symbol of luxury and opulence – in the Old Testament, Amos warns that 'the houses of ivory shall perish'. Elephants were hunted in Egypt 3,000 years ago. By about 2,500 years ago they were extinct in Syria. In AD77 Pliny complained that ivory was only to be had from India '…the demands of luxury having exhausted all those in our part of the world [i.e. most of North Africa]'. This is not so surprising when you consider that Caligula was reputed to have built an ivory stable for his horse. By about AD500 elephants were close to extinct north of the Sahara.

In the rest of Africa the process was slower but the end result was much the same. About 500 years ago Europeans started the slaughter in the rest of Africa which continues today. Thousands of elephants died in the nineteenth century to provide knife handles, piano keys and billiard balls: a full snooker set (22 balls) could account for three or four tuskers; big-game hunters out for a thrill accounted for many more. In the twentieth century the pace has quickened and in last fifty years the remaining herds have been decimated.

Most scientific research into the natural world has been done in the last fifty years or so. Prior to this most knowledge came from apocryphal stories of travellers, adventurers and hunters. In the last few decades an increased number of scientists, technological advances and a new concern for ecological matters has led to a scientific assault upon nature. With advances in technology like super slow-motion cameras, night-vision, motorised and remote sensors, etc., we are able to observe behaviour that until recently was totally hidden. We can monitor bird and animal migrations or listen to whale-song or look inside a termite mound. We can go to the depths of the ocean or into the upper atmosphere or to the middle of Antarctica and see how life behaves in these previously unexplored environments.

But what exactly is it that these scientists are studying? Can one make any observation on the 'natural' behaviour of, say, elephants when what one is looking at is a remnant population confined within Man-made boundaries? We may understand how elephant society works now under these restricted conditions but whether this is 'natural' behaviour which would be the same is the absence of Man is another matter.

I have used the elephant as an example but I could have used the bear, the wolf, the whale, or any of dozens of other animals. The interest in animals as objects worthy of consideration in their own right rather than as a resource to be exploited is only very recent. We have learnt more about the behaviour of whales in the last fifty years than in all our previous history. Prior to that we were too busy killing them to worry about things like whale-song or whether they might have a big brain (unless you could eat it). Only when it ceased to be commercially viable to hunt did we develop a conscience and by that time the damage was done.

World population has quadrupled since the beginning of the twentieth century from about one point five billion to somewhere around six billion when I was younger and it is now approaching, or may even have passed, seven billion. To put this in some sort of perspective there are less than one *million* apes of all other kinds and this number is dropping daily. The long-term picture is less clear. There are some initiatives to curb population but these are inconsistent and unevenly applied and there are many other initiatives which are aimed at easing world hunger, improving health and longevity and reducing disease, all of which will increase numbers. The current forecast is that population will increase further in the short term to at least ten billion.

The majority of people now live in cities and more and more people are moving to urban centres every day. Cities of one or two million are common and there are dozens of cities of over ten million people – that is more than the population of the whole world just 10,000 years ago. We are now so far away from nature that, in the West at least, most peoples' only experience of wild animals is on the National Geographic channel. We wear artificial fibres and eat shrink-wrapped, processed food. We travel from air-conditioned homes in air-conditioned cars to air-conditioned offices. The local park now serves as a token oasis of green in a concrete urban landscape. Nature is not allowed to intrude.

We do not know just how many people the world can bear before something gives, nor do we know what may give or how, but we may be close to finding out. We are perhaps entering a new era – an era where the Earth bites back. The first small symptoms of global warming are showing themselves – storms, floods, wildfires and so on. Weather patterns are no longer so predictable. The monsoon is no longer so reliable. There seems little doubt that this will get much, much more extreme. The sad fact is that resources are finite and we cannot keep on exploiting and expanding indefinitely. The recent huge increase in humanity has resulted in expansion into more and more marginal areas. Natural disasters like floods or tsunamis or volcanic eruptions in isolated areas that a few hundred years ago were sparsely inhabited and would have passed barely noticed, today devastate densely populated cities and result in massive human suffering. We are in an entirely new situation but we are facing it with the same old attitudes.

Whether human technology and innovation can solve the world's problems is a question which we may soon have to answer. Although it is easy enough to allow population to swell to ten billion or more simply by doing nothing to stop it, this will put an enormous burden on already stretched world resources. There are already millions of people living in need and population is rising fastest in many cases in areas least capable of supporting it.

Man is an expansive, destructive species at odds with the rest of nature and there seems little reason to expect that this will ever change. Even if new innovations in genetic engineering allow us to feed the world and alternative energy sources give us the power to run it there seems little doubt that instead of learning the lessons of history and controlling our numbers, we will allow population to expand even further to use up and exhaust any new margin we create.

It seems likely that the first mitochondrial Eve was one of a small number of survivors of an early population squeeze. Perhaps over the long term Man's population goes in a boom/bust cycle where it expands to outstrip resources then crashes and then expands again to overfill the void. Man has been around for so very short a time that predicting the long term future is pure guesswork. If this is the pattern then it could be that the next mitochondrial Eve may have a slightly different genetic mélange than previously and Man may take a new path, another skew in evolution.

Man versus Time

In the end Man is a transient phenomenon the same as any other. His brief flare of existence will fade and disappear in time. Our modern cities will crumble to be re-absorbed into nature in the same way as pre-Columbian cities of some centuries ago. The atmosphere will renew itself and the seas will wash themselves clean. Eventually, in eons to come there will be no obvious sign that Man ever ruled the Earth...except one.

We have sent space craft to the outer reaches of the solar system with the intention that they should sail beyond into the endless depths of the universe. What will be the ultimate fate of this high-tech garbage (and it is ultimately garbage – once it is beyond reach of our communication it is of no use to anybody) does not seem to be of any concern. There are already thousands of pieces of space detritus orbiting the planet. It is a sobering thought that after Man ceases to exist as a species on this planet, after all evidence of our works has been subsumed into the earth, there will still be human artefacts circling above us in the inert stillness of space. Human occupation will be identifiable in the distant future, in the same way as the distant past: by our trash.

How significant Man will turn out to be in the long term it is hard to say. If Man is the agent of the sixth extinction – the one to bring to an end the Age of Mammals – then his influence could be profound and far-reaching. On the other hand if he disappears in the next few hundred years in a nuclear Armageddon or gets wiped out by some stealthy virus then his long-term impact could be negligible.

As I said at the beginning of the book, Man has no nature, what we have instead is culture. Culture is a matter of choice and we can choose to continue on our destructive path or we can choose to change.

What is the future of Man?

It's your choice.

Bibliography

The Procession Of Life: Alfred S Romer -1968 William Clowes & Son
Darwin And The Spirit Of Man: Alister Hardy - 1984 Collins
In The Blink Of An Eye: How Vision Kick-started The Big Bang Of Evolution: Andrew Parker - 2003 The Free Press
Our Cosmic Origins: Armand Delsemme - 1998 Cambridge University Press
Written In Blood: Beverley Macdonald - 2003 Allen & Unwin
The Journey From Eden: Brian M Fagan - 1990 Thames & Hudson
The Seven Daughters Of Eve: Bryan Sykes 2001 Norton
Smithsonian Intimate Guide To Human Origins: Carl Zimmer - 2005 Harper Collins
Seed To Civilization: Charles B. Heiser, jnr
Biology And The Riddle Of Life: Charles Birch - 1999 University of NSW
The Origin Of Species: Charles Darwin
The Spark Of Life: Christopher Wills & Jeffrey Bada - 2000 Perseus Pub'ing
Atlas Of World History Vol.3 The Dark Ages: Colin & Sarah McEvedy
Wildfight: A History Of Conservation: Colin Willock - 1991 Johnathon Cape
The X In Sex: David Bainbridge – 2003 - Harvard University Press
Sociobiology: The Whisperings Within: David Barash – 1979 Harper & Row
The History Of The Pacific islands, Kingdoms Of The Reefs: Deryck Scarr
The Human Zoo: Desmond Morris
The Naked Ape : Desmond Morris - 1969
Prehistoric Textiles: E J W Barber - 1991 Princeton University Press
The Anatomy Of Human Destruction: Erich Fromm - 1973

What Evolution Is: Ernst Mayr   2001 - Basic Books

Our Posthuman Future: Francis Fukuyama - 2002 - Picador

Medieval Civilization: Jaques le Goff

Extinct Species Of The World: Jean-Christophe Balouet - 1990

A Species In Denial: Jeremy Griffith – 2003 - FHA Publishing

Ascent To Civilisation: John Gowlett - 1984

The Evolution Of Culture In Animals: John Tyler Bonner -1980- Princeton University Press

Refuting Evolution: Johnathon Sarafati – 1999 - Answers in Genesis

The Layman's Guide To Infinity: Kenneth W Fraser – 2004 - Zeus Publications

On Life And Living: Konrad Lorenz – 1990 - St Martins Press

The Word And The sword: Leonard M. Dudley

Acquiring Genomes - A Theory Of The Origin Of Species: Lynn Margoulis & Dorion Sagan 2002 Basic Books

Wild Minds: What Animals Really Think: Marc D. Hauser – 2000 - Henry Holt & Co

Elephant Destiny: Martin Meredith - 2001

The Road To Now: Taking Stock Of Evolution & Our Place In The World: Melvin Bolton – 2001 - Allen & Unwin

The Hidden History Of The Human Race: Michael A Cremo & Richard L Thompson – 1996 - Bhaktivedanta Book Publishing

Extinction – Evolution And The End Of Man: Michael Boulter – 2003 - Harper Collins

Intelligence In Animals: Michael Bright - 1994 - Readers Digest

On Becoming Human: Nancy Makepiece Tanner – 1981 - Cambridge University Press

Fossils, The Evolution And Extinction Of Species: Niles Eldredge – 1991-

Where Worlds Collide – The Wallace Line: Penny van Oosterzee – 1997- Reed Books

Power And Greed: A Short History Of The World: Philippe Gigantes – 2002 - Constable & Robinson

Patterns In The Mind: Ray Jackendoff – 1994 - Harper Collins

Plague Species: Reg Morrison – 2003 - New Holland Publishers

A Perfect Harmony: The Intertwining Lives Of Animals & Humans Throughout History: Richard A Caras – 1996 - Simon & Schuster

Climbing Mount Improbable: Richard Dawkins – 1996 - Penguin

Unweaving The Rainbow: Richard Dawkins – 1999 - Penguin

140

The Selfish Gene: Richard Dawkins
The Blind Watchmaker: Richard Dawkins
The Meaning Of It All: Richard Feynman – 1998 - Penguin
African Genesis: Robert Ardrey - 1961
The Territorial Imperative: Robert Ardrey – 1967
The Journey Of Man: Spencer Wells - 2002
World Fire: The Culture Of Fire On Earth: Stephen J Pyne – 1995- Henry Holte & Co
Time's Arrow, Time's Cycle: Stephen Jay Gould – 1987 - Harvard University Press
The Theory Of Everything: The Origin & Fate Of The Universe: Stephen W Hawking – 2002 - New Millenium Press
In The blood - God, Genes & Destiny: Steve Jones – 1996 - Harper Collins
The Nature Of Horses: Steven Budiansky - 1997 - Weidenfeld & Nicholson
Words And Rules: Steven Pinker – 1999 - Weidenfeld & Nicholson
Land Of Lost Monsters: Ted Oakes – 2003 - Hydra Publishing
Human Evolution, Language And Mind: William Noble & Iain Davidson – 1996 - William Noble
From Lucy To Language: Blake Edgar and Donald Johanson – *2006* Simon & Schuster

Printed in Australia
AUOC010809240512
252392AU00001BA/2/P